This book is dedicated to my deadly crew.

To the production staff who labour tirelessly over our vast risk assessments and make everything work like clockwork. To the crew on the road, with all their ideas, enthusiasm and patience; and to my camera and soundmen, constant companions who follow me into the murkiest of swamps and darkest oceans just 'cos I tell them it'll be OK!

Thank you all.

SB

Contents

The Most Lethal Animals on Earth 6

Speed 8

Camouflage and Traps 22

Bite 36

Power 50

Senses 64

Teamwork 76

Extremophiles 84

Endangered 94

Index and Acknowledgments 95

The Most Lethal Animals on Earth

Over the last few years, I've been lucky enough to circle the planet many times as I've searched for lethal beasts, with the aim of capturing them on camera and trying to work out what makes them so special. My adventures have given me some of the most extraordinary experiences of my life, and provided endless excitement and moments of euphoria. In a spooky, abandoned gold mine in Northern Australia I became the first person ever to film carnivorous ghost bats. I've swum with shoals of hungry piranhas and sharks, I've held a lion's paw in my hand, and I've been charged by crocodiles as well as a tiger, an elephant and a hippo.

In all that time, the only nasty injury I've had is a nip from a spectacled caiman – and that was because I stepped on it in the swamps of Argentina. Ten stitches and a great scar is all I have to show for years and years of coming face to face with the animals we consider the most dangerous in the world – proof, if you need it, that these wondrous beasts bear us no malice and are truly more frightened of us than we are of them.

Human beings will always be fascinated by predators, the scintillating creatures that make a living by catching and killing other animals. Our own species is first known to have stood upright and used tools about three million years ago, and for the vast majority of the time since, we have lived surrounded by wild animals. Knowing which animals could harm us, and being interested in how they work, would have given us an evolutionary advantage and kept us alive. Nowadays we live in a very different world, in which animals pose us little or no threat, but we still maintain the vestiges of our caveman past and part of that is our deep-seated desire to know more about predatory animals.

For me, this goes way past the common preoccupation with sharks and snakes. The question I get asked most (after "what is your favourite animal?") is "which is the most deadly animal in the world?" What they usually mean is what is the most deadly to us humans. My answer which often surprises is the *Anopheles* mosquito because of the diseases it spreads. You're more likely to be killed by a falling soft drinks machine than you are by a shark in the modern world. If you're talking about animals in general, though, I think every predator is equally deadly, and equally worthy of consideration. After all, the humble ladybird scoffing down aphids on our rose bushes is just as perfectly adapted to its job of catching its food as tigers are to their role as top predators. In fact, if you look at ladybirds' ratio of successful hunts, their diversity and their numbers, then they are far more successful than tigers!

This book looks at what it is that makes predators excel in their chosen method of hunting.
We go beneath the skin of animals to see how their physiology drives their actions and learn how a creature's skeleton or teeth define the way it goes about its business. We examine how senses allow hunters to track and focus on their food, and discover the super-senses some animals possess that make us humans seem positively bereft. And we also find out how behaviour, teamwork and intelligence can come together to make an organism more than the sum of its parts.

This is the story of the most lethal animals on Earth – the Predators.

Speed

Cheetah

The cheetah can run at up to 112 kilometres an hour. It is the best known speed specialist on the planet, with a body shape that allows it to make longer, faster chases than any other cat. Its head is small, light and streamlined, its bones long and slender, and its heart and lungs tremendously developed. But the chase cannot last longer than about 400 metres – any more, and the cheetah can suffer heat exhaustion and may die. A cheetah usually needs to lie in the shade and pant for at least half an hour after a chase to stabilise its core body temperature.

My race with a cheetah

Who on earth would be dumb enough to try and outsprint a cheetah? Well, err, me! In fact, I wasn't really trying to race Savannah, a cheetah who's used to humans, but trying to get her to show off her top speed. To begin with she looked bored as I tried sprinting off into the distance, but eventually she sensed some fun was to be had, and set off at full pace. She was at top speed in just three strides.

As it runs, the cheetah spends half the time with all its feet off the ground, almost flying. Thanks to its flexible joints and spine, the body stretches out until it is nearly horizontal, and then the spine snaps back under elastic force like an archer's bow.

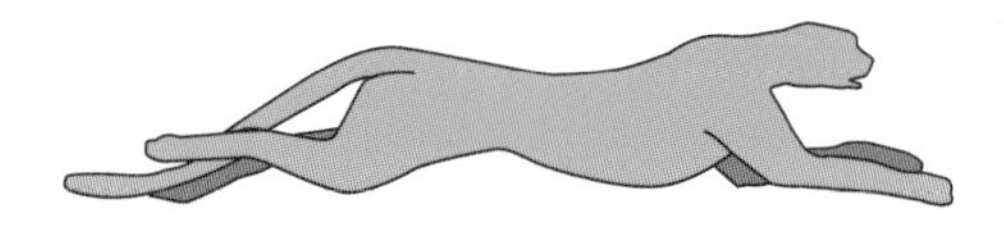

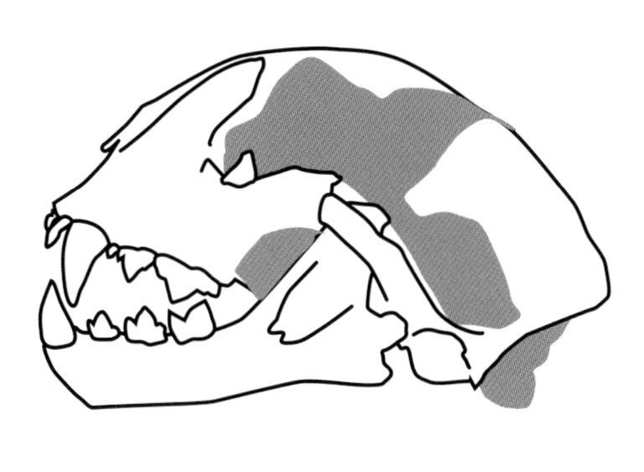

A cheetah's skull is tiny and lightweight in comparison to that of a jaguar or leopard, which are not much larger in body size. This aids the cheetah in its quest to be lean and quick, but also means it is quite fragile. A well-placed kick from even a small

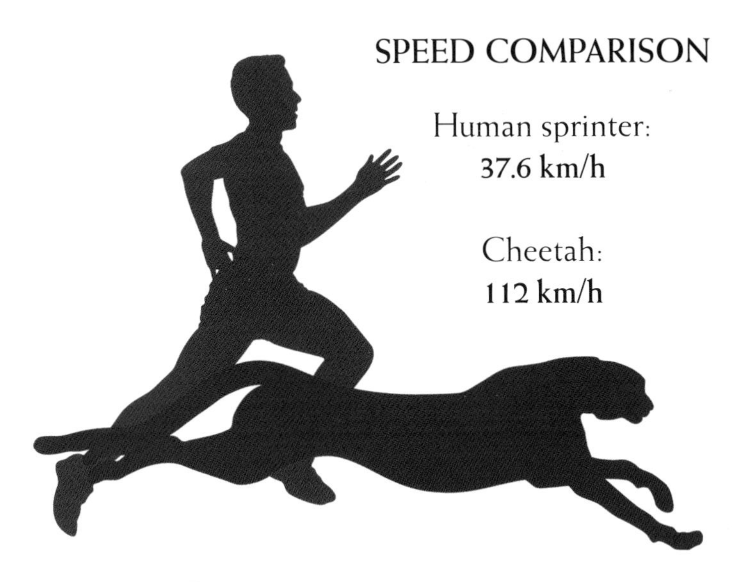

SPEED COMPARISON

Human sprinter:
37.6 km/h

Cheetah:
112 km/h

Speedy Snakes

Snakes might not have legs, but they can move surprisingly fast. Believe it or not, snakes evolved from animals with legs, similar to modern lizards. Over millions of years they lost their legs as they became so good at moving without them – legs just got in the way. Some snakes today still have little bony spurs that show where the pelvis and back legs used to be. Snakes have different ways of moving, depending on where they are and what they are doing.

BLACK MAMBA
This is the longest venomous snake in Africa – up to 4.5 metres long, which is longer than two very tall people lying head to toe. It is said to be the fastest snake in the world, able to move at 20 km/h, but I've run alongside one and kept pace easily. The same cannot be said of snakes such as coachwhips and some racers, which I reckon would be the real record holders.

Chasing a glossy racer

This snake led me a merry dance as it whizzed up bushes, across the ground and through water, barely changing pace in the different environments. Many snakes can move just as fast in water as they do on land. Their sleek outline creates little drag, and they make undulating movements to whip their bodies through the water.

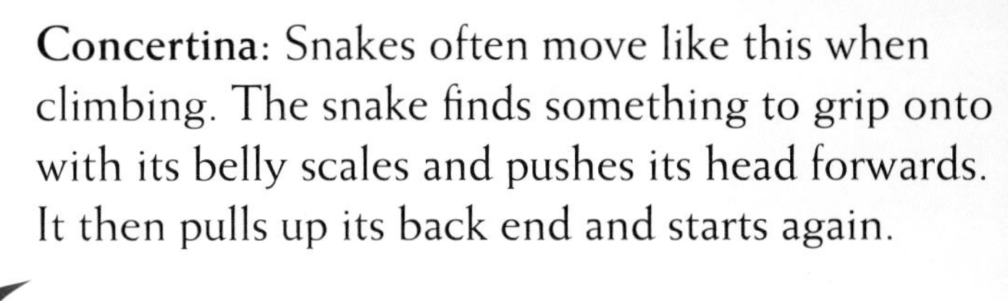

Concertina: Snakes often move like this when climbing. The snake finds something to grip onto with its belly scales and pushes its head forwards. It then pulls up its back end and starts again.

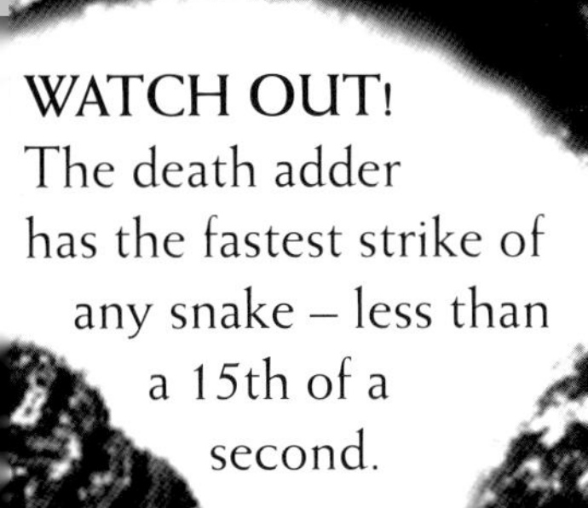

WATCH OUT!
The death adder has the fastest strike of any snake – less than a 15th of a second.

Serpentine: This is the way most snakes move in water, weaving the body from side to side and pushing against the force of the water. On land, the snake moves in a similar way, pressing its body against any little lumps and bumps on the ground to push itself forward.

Sidewinding: This method is used by snakes moving on sand where there's little to push against. The snake throws itself along in sideways waves, with only two parts of its body touching the ground.

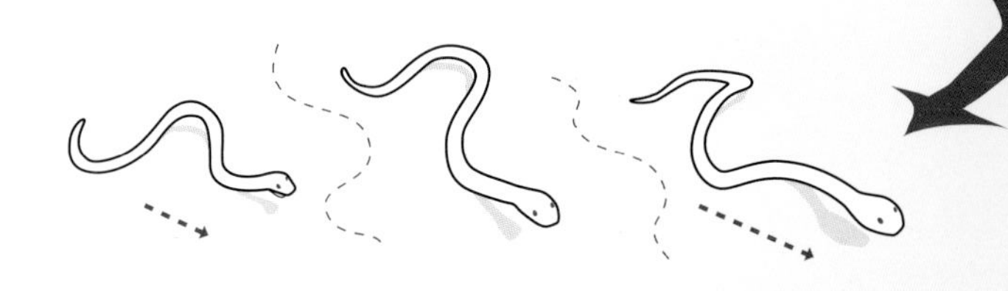

Rectilinear: This is a slow, rippling movement, useful when a snake is stalking prey on the ground and doesn't want to attract attention. The snake uses its belly scales to push against a friction point on the ground and move itself forwards to the next point.

GLOSSY RACER
This snake was nearly too fast for me to catch! It grows up to 1.2 metres long.

Fast Fish

Water is about 800 times denser than air, so moving through it at speed is particularly challenging. To help them move fast, speedy fish have a special hydrodynamic shape – the body tapers at each end to reduce drag and cut through the water. Many fast fish also have oily skin that speeds the movement of water across the body, reducing drag. A shark's skin is covered with thousands of tiny spines called denticles. These create micro-turbulence, which makes the water itself 'oily' and eases the shark's passage through the water.

Searching for the speedsters

Encountering the fastest fish in the sea can be difficult, as they tend to be deep-sea specialists. I've often seen barracuda and tuna at drop-offs where a coral reef meets the deep sea, but marlin and sailfish are harder to spot. We spent two days in the seas of Mozambique dragging an underwater camera behind us to try and entice a marlin in, but unfortunately never struck lucky!

When diving with sharks you are always aware that you have ventured into their world. In my SCUBA gear, I'm slow and sluggish, while sharks scythe through the water with grace and ease.

Thousands of tiny teeth called denticles, shown below in this magnified picture, feel rough like sandpaper. They work so well for sharks that the idea has been copied for the suits of Olympic swimmers.

SPEEDY MARLIN

The marlin is one of the fastest fish in the sea and can move at up to 80 km/h in short bursts. It may use its long pointed snout to thrash out at fish, knocking them senseless before gobbling them up.

SPEED CHAMPIONS

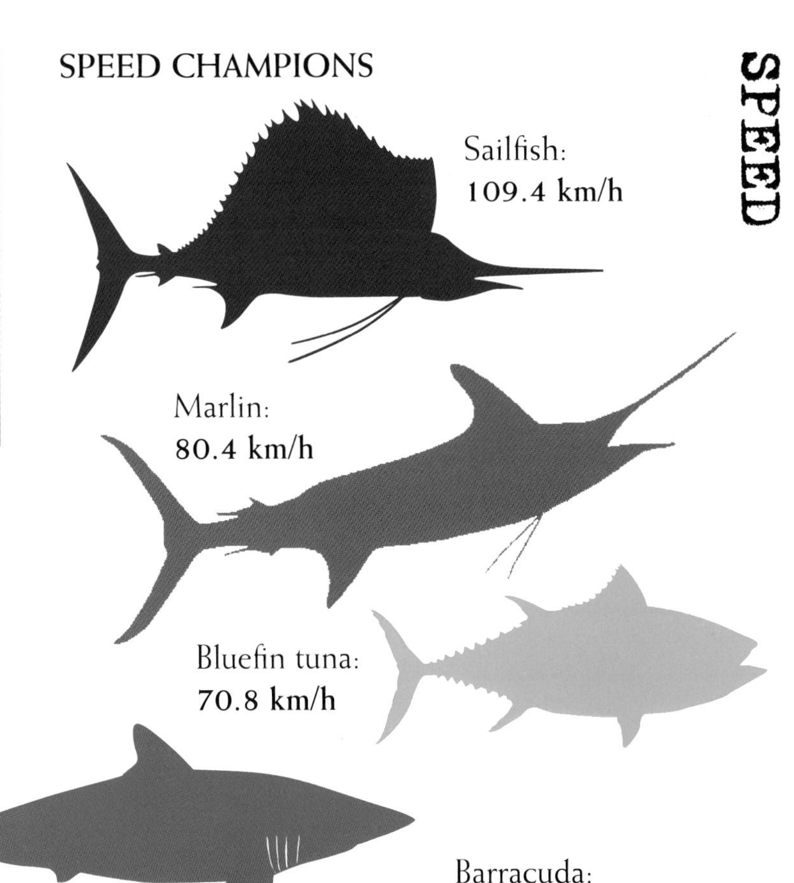

Shortfin mako shark:
69.2 km/h

Barracuda:
58 km/h

Olympic swimmer:
8 km/h

Mouth
The bluefin tuna swims with its mouth open in order to keep oxygen flowing into the body.

Tail provides propulsion so needs to be broad to drive as much water as possible while creating minimum drag. The crescent is the perfect shape for this.

Pectoral (side) fins help keep the fish steady in water but can be folded in when moving at speed.

LONG -DISTANCE SWIMMER

The bluefin tuna has a high proportion of blood-infused and oxygen-demanding red muscle which allows it to swim fast for long distances. These fish swim across the Atlantic in less than two months.

More Sea Speedsters

Marine mammals, such as whales and seals, have their own ways of moving through the water at high speed – very different from fish. Mammals have the benefit of being warm-blooded, which generally means they can maintain high speeds for longer. However, they have the disadvantage of having to return to the surface to breathe air, and also having to possess thick layers of fat or fur to keep them insulated in water and their core body temperature high. They also need to eat frequently to fuel their ravenous metabolisms!

STREAMLINED SEA LIONS

Sea lions push themselves forward with their front flippers and their streamlined body shape reduces drag. A sea lion's bristling whiskers are so sensitive that they can pick up the wake a fish leaves behind in the water as it swims away.

PORPOISING

Penguins, sea lions and dolphins all leap out of the water as they swim – a behaviour known as porpoising. This means they spend some of their travelling time in the air, a less dense medium than water.

My dolphin friends

Marine mammals are often quite easy to interact with underwater. Their intelligence and their inquisitive nature means they'll often come to investigate something new in their world, and they will play with divers, obviously enjoying the company.

DOLPHIN
In contrast to the fast fish, the dolphin has a tail fin that is orientated horizontally, not vertically. It pushes itself forward by moving its body up and down along the length of the spine – more like the motions of a cheetah or a horse than a tuna.

SWIMMING PENGUINS
Penguins swim by using their wings in the same way as sea lions or turtles use their front flippers. This is an example of convergent evolution, which simply means that totally unrelated species develop the same mechanisms to solve a task. Gentoo penguins can put on spurts of impressive speed, and have been clocked at 35 km/h.

DIVING SEALS
True seals swim by making undulating movements of their body and rear flippers. Some can dive to depths of 1,000 metres or more and hold their breath for more than 100 minutes.

Zooming!

It's difficult to measure how fast insects can fly, but some are extraordinarily speedy. The deer bot fly and the death's head hawkmoth are two of the fastest, and the male horsefly has been timed flying at 145 kilometres an hour. For me, though, the hawker dragonflies have no equals in the air. They may fly as fast as 60 kilometres an hour as they zoom around chasing mosquitoes and other flying insects. And by shifting the angle of the wings, rather like a helicopter adjusts its blades, a dragonfly can hover, as well as fly up, down and even backwards.

Wings might look fragile but they are strong and can be beaten at great speeds to whizz the dragonfly through the air when it's hunting prey.

Long front legs are used to snatch prey.

Claw-like jaws can easily tear apart insect prey. Don't worry, though, the stories about these 'devil's darning needles' biting humans are just myths!

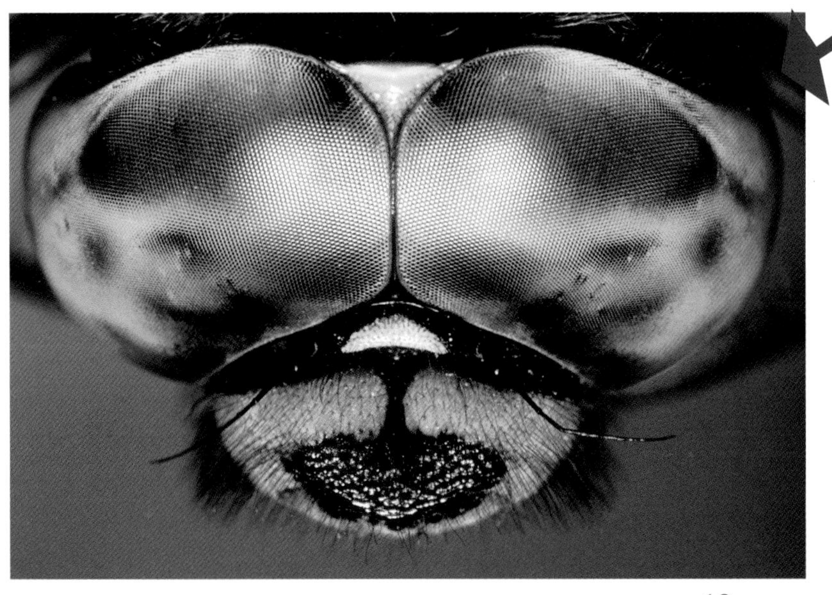

AMAZING EYES

A dragonfly's eyes are very different from ours. They are called compound eyes and are made up of about 30,000 tiny lenses. A dragonfly can see behind itself as well as in front.

Insect eyes are very good at detecting movement. They also have much faster reaction times than we do, which explains why it is so hard to swat a fly!

If you see a dragonfly vibrating its wings while at rest, it is warming up its flight muscles in preparation for the first flight of the day.

Thorax contains the powerful muscles that drive the wings.

The pterostigma (ter-ro-stig-ma) is a tiny counterweight towards the end of the wing, making each wing stroke more efficient.

Speedy solifugids

I took a solifugid on my hand in Mozambique. Although solifugids are non-venomous, they have one of the most fearsome bites of any arachnid, and I have to admit to a bit of a tingle down the spine as it sat there. It didn't bite me, though getting it to sit still for a few seconds was a challenge!

Up close with a dragonfly

One day I saw a dragonfly that had only just completed its metamorphosis and emerged from its nymphal skin to become an adult. It was sitting on a stalk by the water, waiting for its wings to strengthen. This is the most vulnerable time for this normally fearless predator, but it sat on my hand, allowing me the rare treat of feeling a dragonfly vibrating its wings.

RAPID ARACHNID
The solifugid or wind scorpion, a relative of the spiders, may be the fastest running of all invertebrates. Scientists have actually timed these terrifying-looking beasts running on miniature treadmills at more than 16 km/h.

Special tactile hairs on the body and legs help a solifugid sense movements in the air and on the ground so it can home in on prey with great accuracy.

Peregrine Falcon

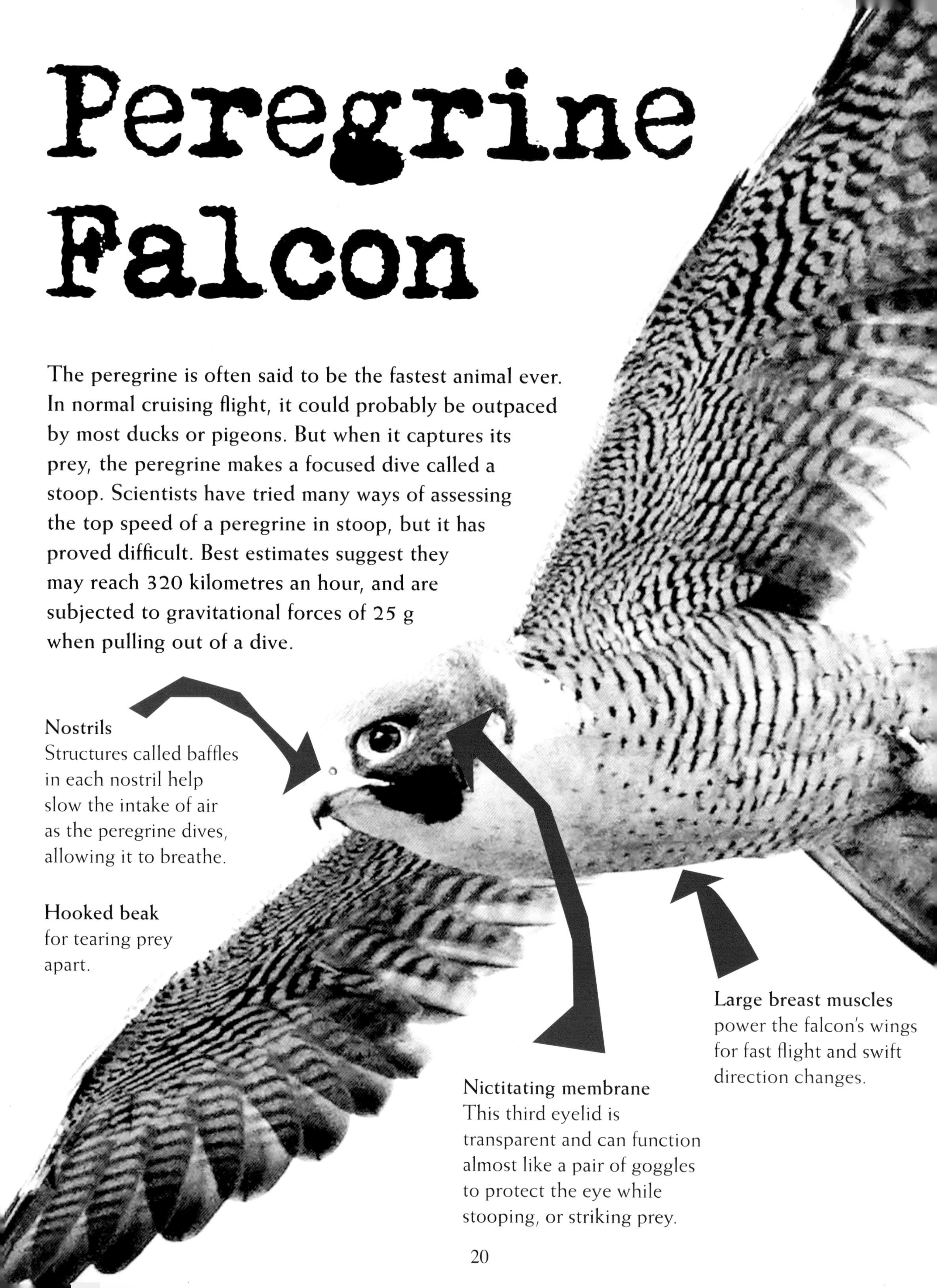

The peregrine is often said to be the fastest animal ever. In normal cruising flight, it could probably be outpaced by most ducks or pigeons. But when it captures its prey, the peregrine makes a focused dive called a stoop. Scientists have tried many ways of assessing the top speed of a peregrine in stoop, but it has proved difficult. Best estimates suggest they may reach 320 kilometres an hour, and are subjected to gravitational forces of 25 g when pulling out of a dive.

Nostrils
Structures called baffles in each nostril help slow the intake of air as the peregrine dives, allowing it to breathe.

Hooked beak
for tearing prey apart.

Large breast muscles
power the falcon's wings for fast flight and swift direction changes.

Nictitating membrane
This third eyelid is transparent and can function almost like a pair of goggles to protect the eye while stooping, or striking prey.

Wings are held back to streamline the bird's shape when it dives.

Curved talons with pointed tips.

BAR-HEADED GEESE

I'm intrigued by migrating bar-headed geese, which travel over the Himalayas at altitudes of over 9,000 metres above sea level. This is way up in the high velocity winds known as the jetstream, which can rage at more than 320 km/h. Though it has never been measured, it is surely possible that the bar-headed goose could use these winds to cruise at the same speed as a peregrine stoops!

Bar-headed geese fly higher than Mount Everest.

Speedy bird

We've done many experiments to analyse the true speed of the peregrine. I've been chased by one in a sports car and down a zipline, and I've even taken to the skies myself in a jet fighter to experience a tiny taste of the g-forces a peregrine must endure. I have huge respect for this perfectly engineered predator, which easily outdoes anything human technology can muster!

Cam

uflage
and
Traps

Cunning Cats

Green is the camouflage colour of choice for many smaller creatures, such as insects, frogs and lizards, but not for large mammals. They have too much surface area to rely on simple blocks of green to merge with their background. Instead, mammals break up their outlines making it hard for prey, or other predators, to see them. Their coats have spots, stripes, dots and rosettes, which blend with their surroundings or mimic the play of light and dark on the undergrowth. A tiger's bright stripes look obvious out in the open, but when you see one in its natural surroundings it's clear why a tiger is among the hardest of all animals to find and film!

The tiger is the biggest of the big cats and depends on being able to creep close to prey and make a swift kill. It's too heavy to run fast for long.

CLOUDED LEOPARD
The scientific name of this cat is *nebulosis*, Latin for cloudy. This refers both to the cloud-like patterns on its fur and to its habit of living high up, near the clouds.

A jaguar's rosette markings help keep it hidden in the dappled light of the rainforest floor.

A tiger's stripy markings merge with long grass and bushes and make it hard to see.

A lion's golden coat blends with the dry grass of its savanna home and helps it hide from prey.

Meeting a big cat

I've tracked clouded leopards many times, and found their paw prints and their scat, but the only one I've ever seen was this youngster in a captive-breeding centre in Thailand. This is truly one of the world's most entrancing creatures, though I'd have been much more nervous being in a cage with a full-grown male.

A tiger is up to 3 metres long, including its tail, and weighs more than three adult people. Each tiger's stripes are unique. No two have exactly the same pattern.

Mantids

Anyone who thinks the fiercest predators on Earth are animals like lions and sharks should take a closer look at praying mantids. They lock onto a moving target with acute vision, then strike so fast that the human eye cannot fully register it. Their range of camouflage is truly awe-inspiring – they take on the form and colour of dead or flourishing leaves, even of the most exotic orchids. They mostly feed on insects, but have been known to catch lizards and small mammals, and I've seen them striking at birds.

DEAD
LEAF MANTIS
This mantis enhances its camouflage by swaying lightly, like a leaf trembling in the breeze.

Eyesight
The mantid's compound eyes are made of thousands of tiny lenses which give it powerful binocular vision. Three simple eyes between the compound eyes are called ocelli, and can probably only tell light from dark.

ORCHID MANTIS
This mantis hides at the centre of an orchid flower and has a steady supply of food as insects come to seek nectar.

Strong mouthparts
The mouth is a complex collection of different parts that pass food into the mouth and break it apart.

Field of vision

Mantids have a dramatically flexible neck, and the triangular shape of the head means more of the eyeball is exposed to the outside world, allowing a vast field of vision.

ANT MIMICS

Most young mantids are tiny wingless versions of their parents. But when they first hatch some look more like ants. Ants can give nasty bites and stings and usually taste bad, so many predators leave them alone. The young mantis gains protection from its ant-like disguise and trickery.

Mantis feast

In Bhutan, I watched a female mantis laying her egg case. She spun a sort of liquid candy floss from the end of her abdomen, which hardened into sponge to protect the eggs. Just seconds after she'd finished, parasitic wasps landed and laid their own eggs into the egg case. The wasp maggots hatch to find a ready-made dinner – they eat up the young mantids.

Chameleons and Geckos

Chameleons are the world's most famous colour changers, though squid, cuttlefish and octopus transform their colours much faster and with even more lurid results! Chameleons can change to match their environment, but their most rapid and dramatic colour changes are in response to mood. When facing off a rival, a chameleon may take on dark, threatening shades, and when courting, it may show off its most gaudy flamboyant tints. Some species will also change colour to control their body temperature, turning dark when they need to soak up sunlight, and paler to reflect the sun's rays and cool down.

Eager feeder

The chameleon's sticky-tipped tongue is a spectacular weapon. It can be longer than the chameleon's body, and can be fired out to its full length in a 16th of a second – fast enough to catch a fly or to snatch a cricket from my fingers!

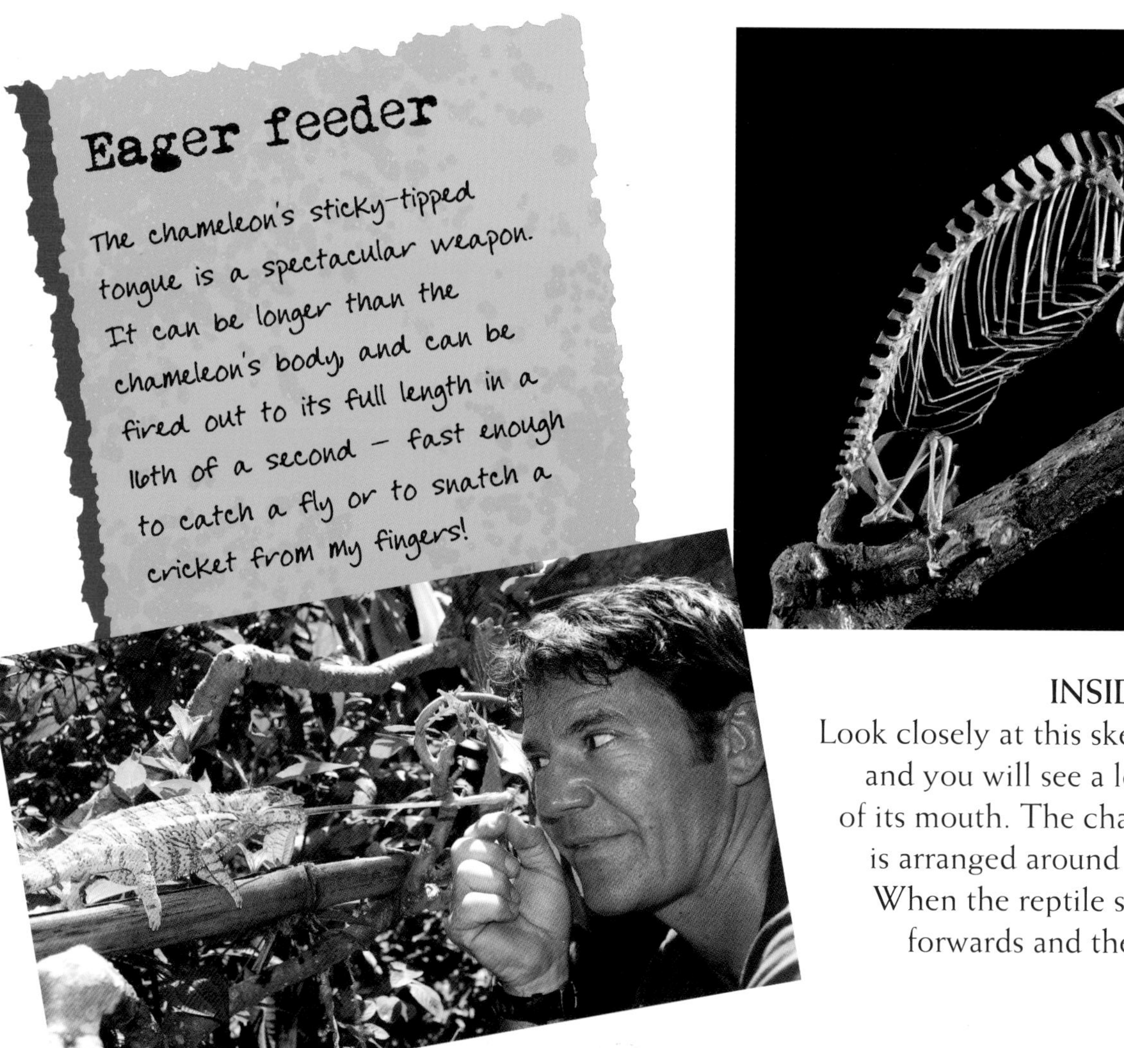

INSIDE A CHAMELEON
Look closely at this skeleton of a chameleon and you will see a long bone sticking out of its mouth. The chameleon's long tongue is arranged around this supporting bone. When the reptile strikes, the bone rocks forwards and the tongue shoots along it at great speed.

Swivelling eyes
The chameleon can move its eyes separately so it can look at two different things at the same time!

Gripping feet
The chameleon's feet are arranged with two clawed toes on one side and three on the other, perfect for gripping small branches. The palms are ridged to help grip too.

Strong tail
The tail can be held coiled like this, or be used to grip onto branches.

WHERE'S THE GECKO?

Can you spot the leaf-tailed gecko among the dead leaves? It has a twisted leaf-like body, vein markings and even edges that look as if they have been nibbled by birds.

On one hand, I have the panther chameleon, which can be 50 cm long, and on my finger is the pygmy, which can sit comfortably on a matchstick.

Reef Monsters

There is no more colourful environment than a tropical coral reef, with lots of brightly patterned animals. However, there are also countless camouflaged beasts lurking on the sands or among the corals and plants, that are not so easy to see. Even the fiercest of predators relies on its cryptic colours, not just to creep up on prey but also to avoid the attention of even bigger mouths and teeth. In the sea there's always something trying to eat you!

LEAFY SEA DRAGON

This amazing creature is a fish! The leafy sea dragon (right and below) looks so like a piece of floating seaweed it is almost impossible for predators and prey to spot.

Leafy sea dragons are like swimming vacuum cleaners. They have no teeth, but suck small prey into their snouts with an audible 'pop' sound!

TASSELLED WOBBEGONG
Believe it or not this is not an old rug lying on the seabed but a shark! Its colour is so similar to its surroundings that it is very hard to see. Unsuspecting small fish and shellfish come to investigate the fringes around its mouth, mistaking them for bits of food. When they are within critical distance, the wobbegong strikes with blinding speed.

Face to face with a frogfish

I found this frogfish on the struts of an old oil rig in Borneo. The frogfish hunts by creating a vacuum in the back of its huge mouth, then opening its mouth and lunging forward with great speed to suck fish into its gullet. The fish's stomach can distend to double its normal size in order to take in extra-large food items.

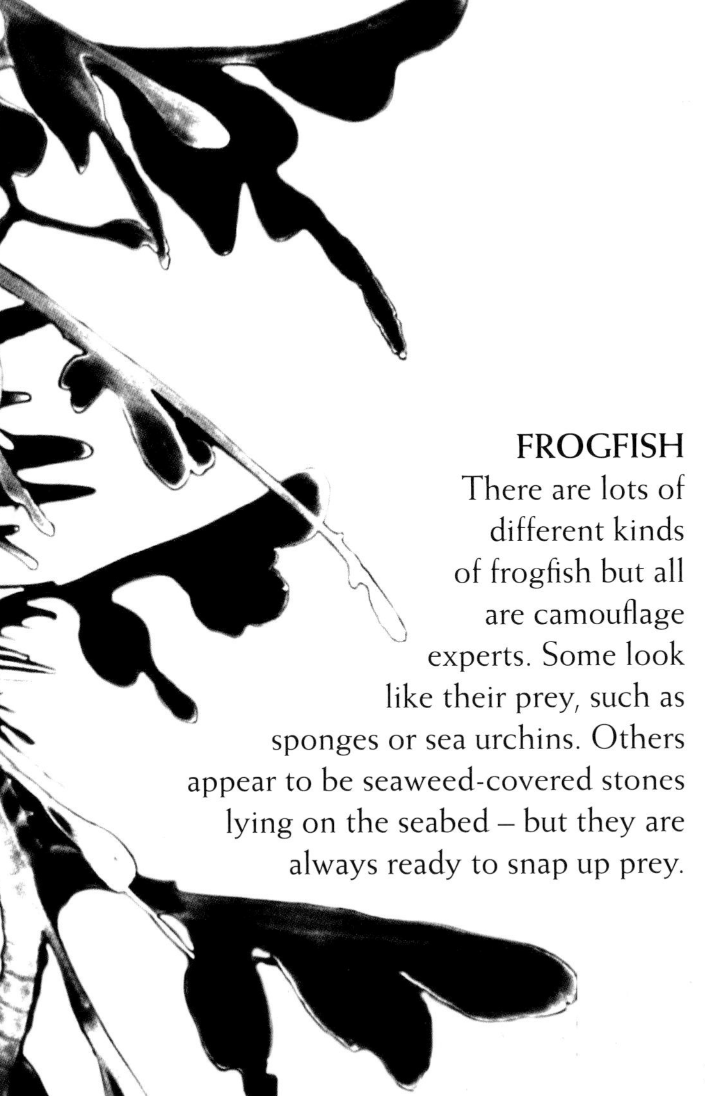

FROGFISH
There are lots of different kinds of frogfish but all are camouflage experts. Some look like their prey, such as sponges or sea urchins. Others appear to be seaweed-covered stones lying on the seabed – but they are always ready to snap up prey.

Mimics

Mimicry in the animal world means the process of one kind of animal taking on the appearance of another for the advantages this could bring. The animal may also take on the scent, calls or behaviour of the other creature in order to hunt more successfully or avoid being caught itself. Some animals mimic others that may be venomous or nasty tasting to predators – even though they are harmless. Their resemblance to the deadly look-a-likes helps to protect them from enemies.

Dangerous or not?

When you see a red, yellow and black snake in the undergrowth, you have to be very, very sure about what you're doing. The coral snakes they mimic are in the cobra family, and are lethally venomous. The false coral snake below is another harmless mimic.

CORAL SNAKE AND KING SNAKE
The coral snake (below) has a powerful venom. The scarlet king snake (above) does not, but when snake-eating predators see the black, red and yellow markings, they avoid the king snake – just in case!

Ant mimics

Ants bite and sting, so few creatures bother to try to eat them. Because of this, more animals mimic ants than any other insect.

This looks just like an ant but in fact it's a spider!

Wasp mimics

Predators see the stripes and assume this is a stinging wasp. In fact, it's a harmless moth, protected by another creature's warning colouration.

The coral snake has a small mouth and short fangs so has to have a good long chew on its victim in order to inject its venom.

MIMIC OCTOPUS

All these pictures are of the extraordinary mimic octopus, which was discovered in 1998 off the coast of Sulawesi in Indonesia. Like its relatives, it has eight arms, but this octopus can twist its arms and body to take on the body shapes of about 15 different animals.

This skilful mimic can take on the shape of a deadly poisonous sea snake or a lionfish. It can flatten its body and swim across the seabed like a flatfish. And it can look just like a jellyfish. The octopus can also change colour, turning its stripes to threatening black.

Spider Traps

Spiders are among the most complex, exciting and bizarre groups of creatures on the planet, with 35–50,000 species, and use a dazzling array of techniques and traps to snag their insect prey. They are certainly the most important predators of insects in the world, and one of their most effective weapons is silk, which a spider makes in its own body. An orb-weaver spider may produce eight different sorts of silk, from the toughest thread used to make the structure of the web, through to the shroud-like silk it uses to truss up its victims. The average web line is only 0.003 millimetres wide, but is said to be five times stronger than steel. It is truly one of the great wonders of the natural world.

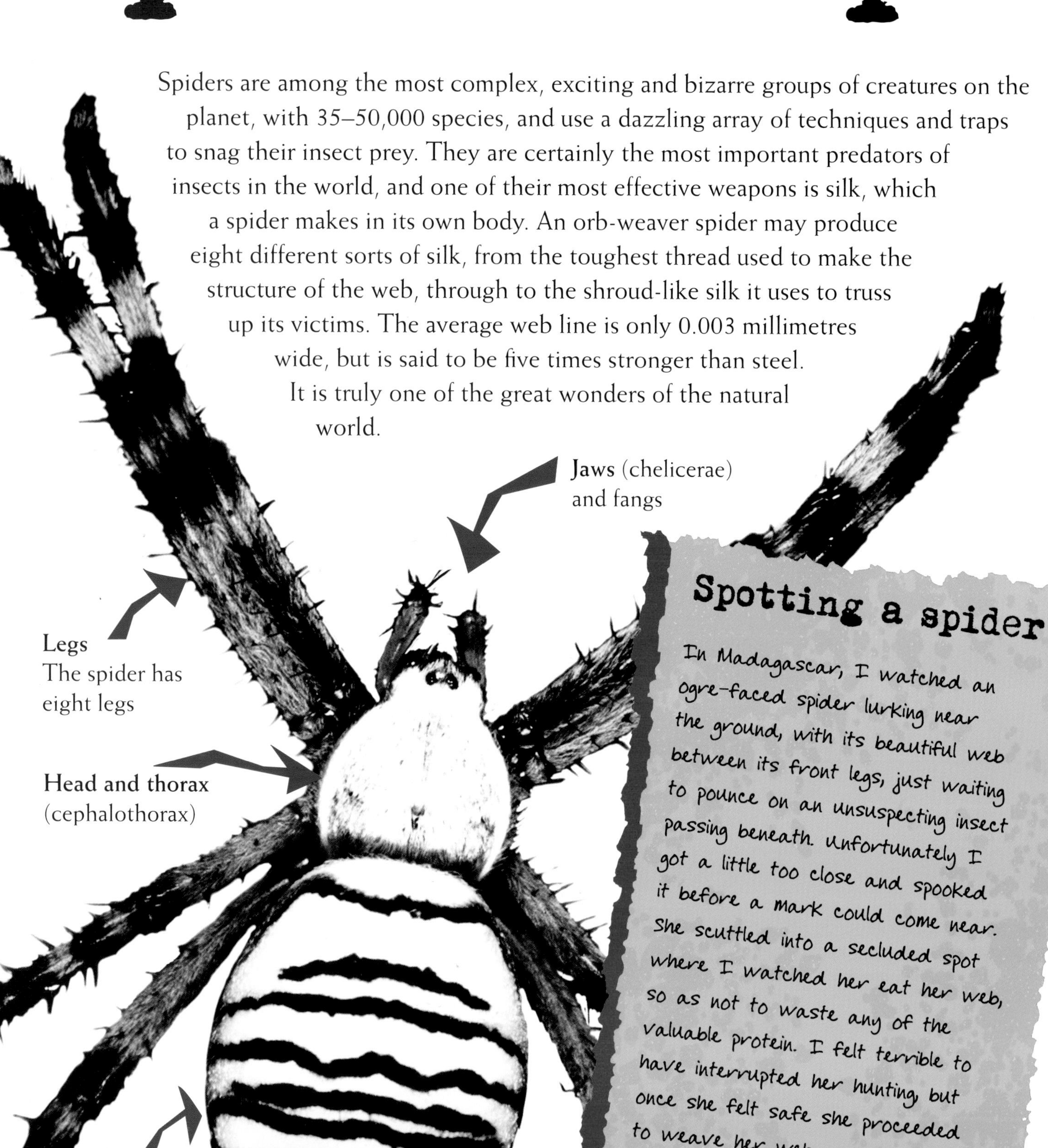

Spotting a spider

In Madagascar, I watched an ogre-faced spider lurking near the ground, with its beautiful web between its front legs, just waiting to pounce on an unsuspecting insect passing beneath. Unfortunately I got a little too close and spooked it before a mark could come near. She scuttled into a secluded spot where I watched her eat her web, so as not to waste any of the valuable protein. I felt terrible to have interrupted her hunting, but once she felt safe she proceeded to weave her web again in a matter of minutes.

TRAPDOOR SPIDER

This large hairy spider digs a burrow and makes a lid out of silk, soil and other material. It leaves triplines like spokes outside the burrow. The spider hides inside, ready to pop out in an instant if anything edible wanders past.

A spider makes silk in special glands in its body. The silk is then squeezed out through little nozzles called spinnerets at the end of the spider's abdomen. Most spiders have two or three pairs of spinnerets.

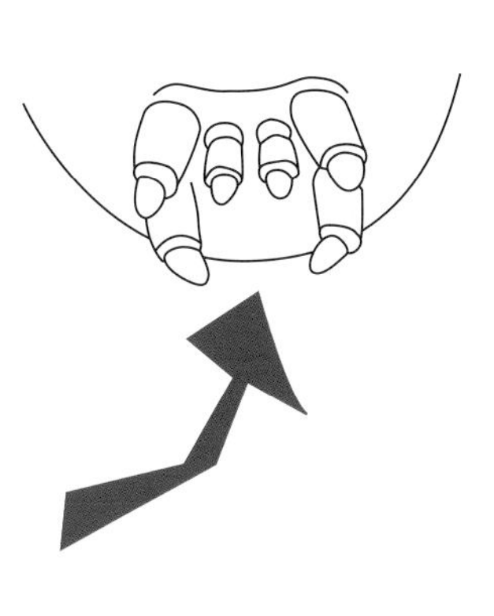

SPINNING A WEB

It took me and my team three days and several thousand metres of rope to build my web. An orb-weaver spider can do it in less than an hour.

First the frame is put in place, with links to surrounding supports.

Then the spider adds more spokes.

Finally it adds a spiral of sticky silk at the centre to trap the prey.

OGRE-FACED SPIDER

The ogre-faced spider has excellent eyesight. It hangs close to the ground with its silken net held between its two front legs. When it spies prey, it pounces, entangling its victim in its net.

Bite

King Cobra

With a record length of 5.7 metres – longer than an average car – the king cobra is easily the longest venomous snake in the world. It has unusually high venom yield, delivering as much as a tablespoon of neurotoxin into its prey, and its fangs are long enough to inject the venom deep into the prey's muscles. In theory, the cobra has enough venom to kill ten humans, or one elephant, with a strike that can prove fatal in a minute and a half. However, kings are actually shy snakes of the deep forest, and will do almost anything to avoid confrontations with people.

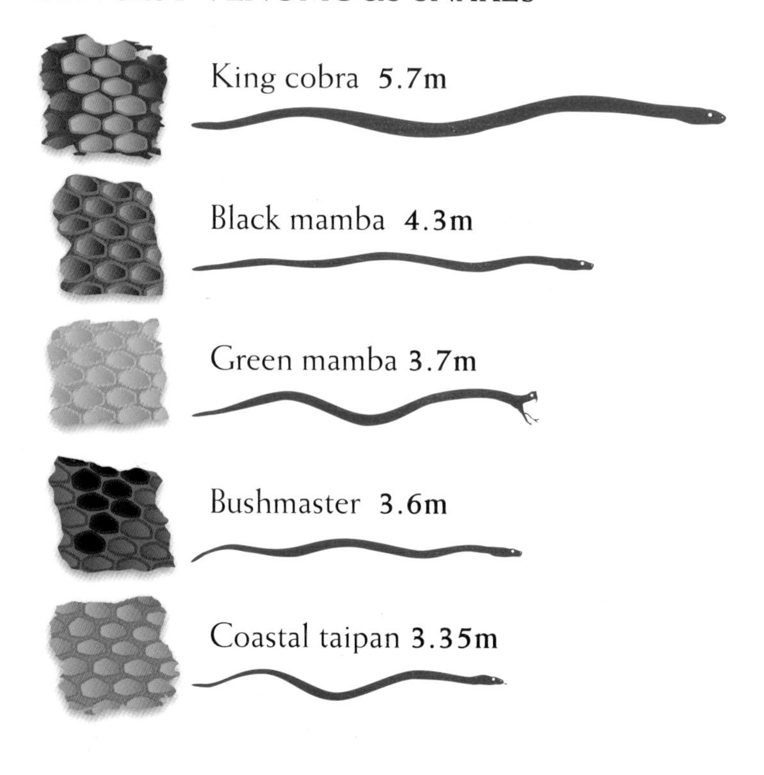

Elapid (cobra) fangs are fixed and hollow. They are comparatively short for the size of the snake.

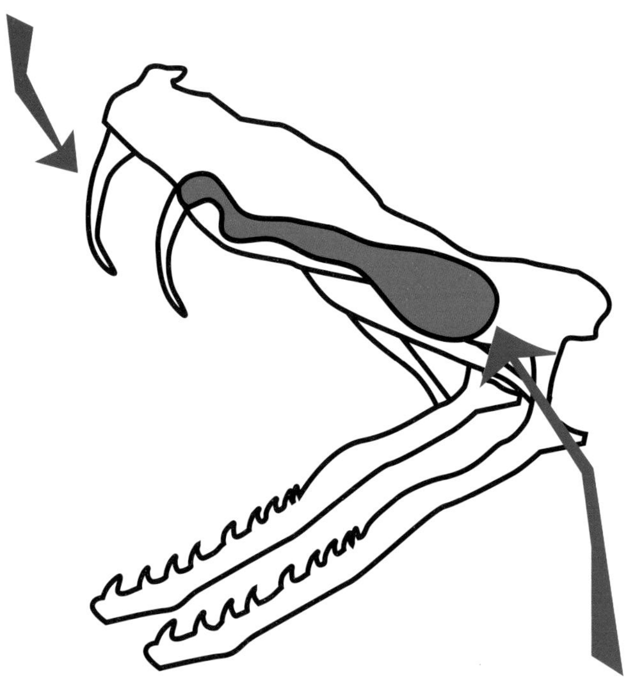

Cobra venom
Venom glands are particularly big in this species. The venom itself is secreted by modified salivary glands, but many of the constituents are created in other organs in the body.

JUNGLE MASTER

The bushmaster is the longest viper in the world. It lives in the jungles of South America and lies half hidden on the forest floor, ready to strike any passing prey.

Unlike the vipers, which have hinged fangs, a cobra's fangs are fixed in position. This means a cobra has to strike downwards from an elevated position. A viper however can strike outwards from a coiled start. A king cobra can hold almost a third of its length erect, so theoretically can look a fully grown person in the eye! It has special membranes in the throat that turn its hiss into more of a growl. Very scary!

Ophiophagus (oh-fee-oh-fay-gus) – the scientific name of the king cobra – means 'snake eater'. This is because kings feed almost exclusively on other snakes.

Catching a cobra

I've been lucky enough to work with King cobras of more than 4 metres long, and they are the most intimidating serpents on the planet. The scariest encounter I've had with any snake was with a sizeable King cobra up a tree. I grabbed hold of its tail, but as it turned round towards me, I realised I had to use my other hand to hold onto a branch and stop myself falling. All of a sudden, catching this particular snake did not seem like such a good idea!

Venom

Strictly speaking,venom and poison are not the same thing. A poison is ingested, eaten or absorbed into the body, while a venom is injected and enters the bloodstream by the use of fangs, teeth, spurs, claws or spines. The pufferfish, for instance, contains a poison called tetrodotoxin in its liver and skin and can be deadly to anything that eats it. The blue-ringed octopus introduces the same substance into its victims with a bite from its sharp beak. The toxin is contained in its saliva (spit). In snakes venom is modified saliva, stored and delivered by highly developed salivary glands.

My bullet ant ritual

Some years ago, I was making a programme about venoms, in which I joined the Satere Mawe tribe of Brazil in their initiation ceremony. This ritual, which marked the passage of boy to man, involved being stung by hundreds of bullet ants. The ants – named bullet ants because any single sting is like being shot – are sewn into palm gloves, with the stingers on the inside. The wearer must keep the gloves on for about ten minutes, being stung repeatedly. To begin with the pain was severe, but perhaps half an hour later as the neurotoxins took effect, the pain totally overwhelmed me, until there was nothing in my world but pain. This continued for perhaps three hours, and the less severe pain for another day or so. Not bad for a little ant!

The sharp sting at the end of the scorpion's body is linked to a venom gland.

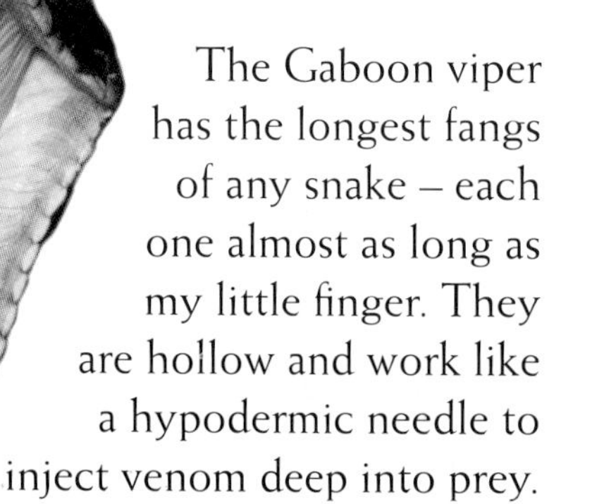

The Gaboon viper has the longest fangs of any snake – each one almost as long as my little finger. They are hollow and work like a hypodermic needle to inject venom deep into prey.

DANGER IN THE SEA

The lionfish uses the sharp spines on its back to deliver its venom into any creature it considers a threat.

When snorkelling in Palau in Micronesia, I hit my ankle on a fire coral, which actually broke off in my ankle bone! Just thirty seconds later I couldn't move my leg at all, and had to spend the night in hospital.

VITAL VENOMS

Russell's viper: **0.75mg**

Giant scolopendra centipede: **1.5mg**

Inland taipan snake: **0.025mg**

Sydney funnel web spider: **0.16mg**

These are the most venomous creatures in different groups and NOT a top ten. If it was, all the top ten venoms would come from snakes. The figures show the amount of toxin in milligrams that is required to kill half the members of a tested population of mice. By the way, the toxin of the dart frog is a poison and is eaten rather than injected.

Lionfish: **4.4mg**

Gila monster: **0.4mg**

Poison dart frog: **0.0008mg**

Stonefish: **0.4mg**

Dubois sea snake: **0.044mg**

Deathstalker scorpion: **0.16mg**

Bite Power

There is relatively little good data for the bite force of wild animals. I have a bite pressure gauge, which has recorded a saltwater crocodile as having an extraordinary 3,500 pounds per square inch of pressure in its bite. But the same test on a great white came up with pathetic measurements – perhaps something to do with the floating nature of their cartilaginous jaw, or because it is so difficult to encourage them to deliver a killer bite to a machine! Lions and tigers only measured 70 or 80 pounds per square inch, before their mighty canines punched straight through the machine!

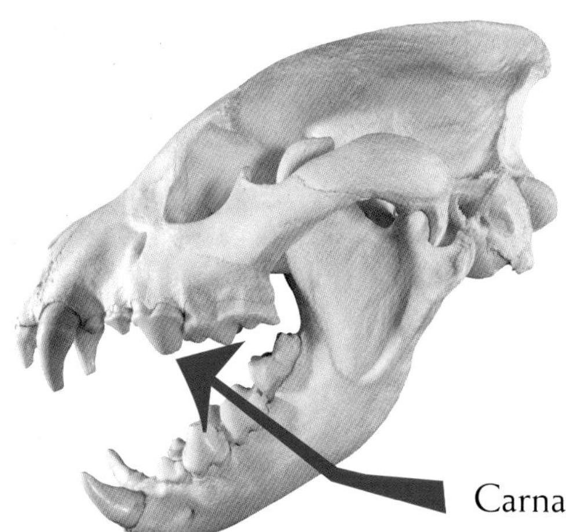

SPOTTED HYENAS

These predators have the ability to crunch clear through bones, using the mighty carnassials or cheek teeth. Their digestive system is so powerful that it can then break down even hooves and horn, making them more efficient than any other carnivore at clearing a carcass.

Monster turtle!

The alligator snapping turtle is known for having one of the fiercest bites of any animal on the planet. Unbelievably, this monster turtle I'm holding is less than quarter the weight of the largest ever-recorded snapper. Apparently the biggest snappers are capable of biting through a broomstick handle!

Glad this is a skull and not the real deal!

The muscles that close a crocodile's jaws are thick and extremely strong. They act with phenomenal speed and crushing weight.

Alligator snapping turtles are among the largest freshwater turtles in the world and can weigh more than 100 kilos.

Reaching 6 metres or more, the saltwater crocodile is the largest of its family.

Terrible Teeth

It's difficult to decide which animal deserves the title of best bite. How do you judge between the mighty force of the saltwater crocodile and the scary-sharp teeth of the leech or the vampire bat? However, the real point is that an animal's dentition is pretty much perfect for its specific food source. The shape of an animal's teeth tells us lots about what it feeds on and how it eats. For example, in the sloth bear which feeds mostly on termites, the premolars are flat grinding plates for mushing up soft prey. In polar bears, which feed exclusively on meat, these teeth are pointed and sharp, like the teeth you'd find in the mouth of a cat or dog.

MONITOR MAWS

Monitor lizards are exceptional swimmers, climbers and sprinters. They eat anything from insects to carrion, and they hunt quite sizeable prey. The largest monitor, the Komodo dragon (above), will kill animals as large as water buffalo! Monitor lizards' teeth are as sharp as a carving knife and can deliver a vicious bite. They've even been discovered to have venom glands, and administer a mix of venom and nasty bacteria with every bite.

BONES AND TEETH COUNT

	BONES	TEETH
Human	206	32
Cat	244	30
Crocodile	250	60-80
Dog	320	42
Snakes	As many as 800	At least 100

We'd been out on the river fishing for a particular kind of catfish for several hours, when we pulled this monster out of the depths. It's called a sabre-toothed kerasin. It has special holes in its skull that function like sheaths, preventing those gargantuan teeth penetrating into the brain.

The living Dracula

The vampire bat is said to have the sharpest teeth of any creature, and I can certainly vouch for their scalpel-like qualities! The first time I caught a vampire bat in a mist net, it turned right around and bit my thumb to the bone, clean through a leather glove. Normally I would have been worried about rabies, but luckily we were in the Amazon rainforest, far away from civilisation and all the domestic animals that can pass on this disease to the vampire bat.

Tiger Shark

The teeth of sharks and rays vary dramatically. There are the crushing plates found in a stingray's jaws that act like grindstones for breaking up shellfish, and the long thin stilettos with lateral cusps that sand tiger sharks use to snag darting prey. Then there are the highly specialised but simple dagger-like teeth of the shortfin mako, compared to the nine different types of teeth of cat sharks, with their multiple cusps and many functions. Even more remarkable is that if you look at a shark's jaw from behind, you can see there are rows of teeth all stacked up and ready to roll into the place of the original once it's been shed.

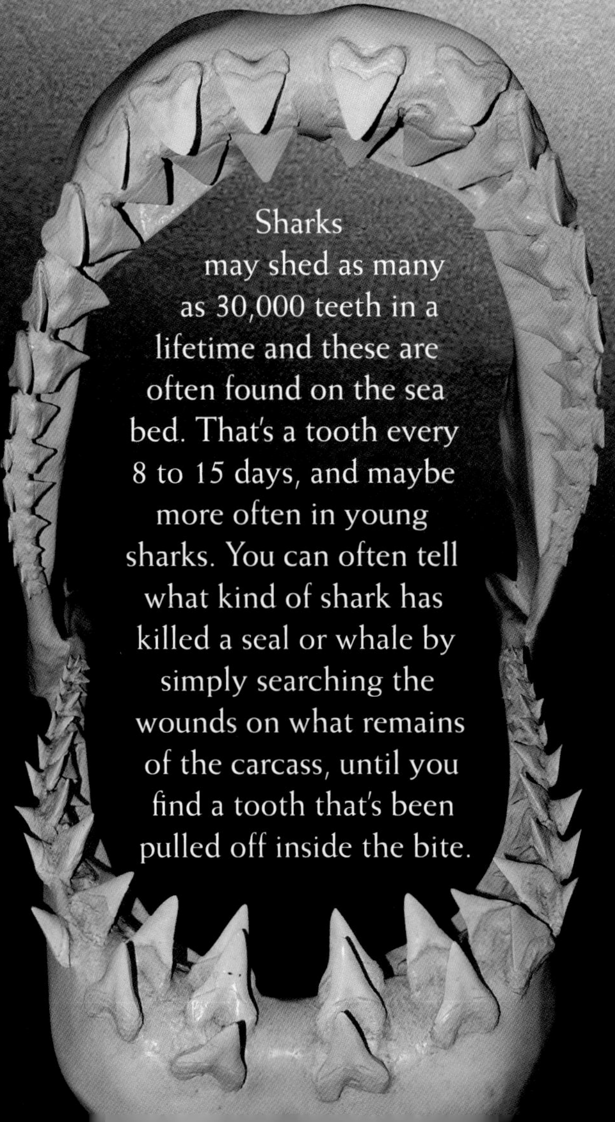

Sharks may shed as many as 30,000 teeth in a lifetime and these are often found on the sea bed. That's a tooth every 8 to 15 days, and maybe more often in young sharks. You can often tell what kind of shark has killed a seal or whale by simply searching the wounds on what remains of the carcass, until you find a tooth that's been pulled off inside the bite.

Great hammerhead shark has triangular teeth with finely serrated edges for cutting into flesh.

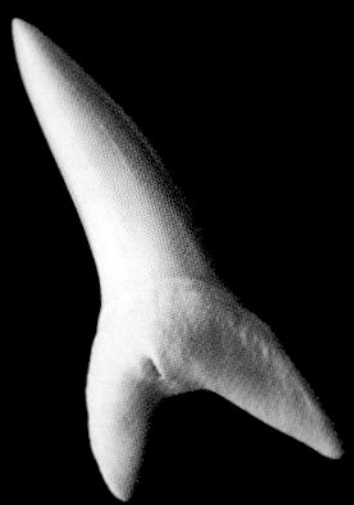

Shortfin mako has long sharp teeth for grasping slippery fish.

Swimming with a tiger shark

I've probably spent more time in the water with tiger sharks than any other kind of shark, and have so much respect for them. Despite their bulk and menacing appearance, they are cautious and careful, and will swim in circles around you at a respectable distance, clearly checking out if you are a threat or not. They have a noted liking for carrion, and will approach food nervously, and are easily spooked away. Once they have built up confidence though, they show off their fearsome strength and power, thrashing away at bait and using their whole body to wrench it around.

Tiger shark's jagged-edged teeth can be used like can openers for attacking hard-shelled prey.

Piranha

Piranhas have a reputation as ferocious fish, which attack in brutal packs. However, this is not normal behaviour for piranhas and I have swum in piranha-infested waters without a single bite – so far! Many species are fruit, seed or leaf feeders, and most will usually hunt alone, sneaking a bite out of another fish then making a dash for safety! Swimming in shoals evolved as a defence against other predators – in a shoal there are many watchful piranha eyes looking out for hunters such as river dolphins, giant otters, cormorants and caimans. However, if the conditions are right, a conveyer belt of piranhas can attack together in a true feeding frenzy, and then the results really do mirror the Hollywood representations.

The red-bellied piranha's blunt, short body is perfect for agile movement. It can zip into food, then turn back on itself extremely quickly and swim away to let the next fish come in for a bite. The broad snout is mostly due to the large jaw muscles that give the piranha such a fearsome bite.

PIRANHA POWER

To prove that piranhas are careful and choosy about what they eat, I stood up to my neck in an Argentinian lake, with a piece of steak held in front of me. It took 15 minutes for them to get up the courage to risk a bite, then we filmed them churning the water white as they scoffed it down. I wasn't bitten once, though I have to admit to being a little nervous!

Piranhas are alert for the sound of splashing or the scent of blood in the water – sure signs that a meal is near.

On an expedition in Guyana we had nets in the entrance to a stream to try and discover new species of fish. Unfortunately a net got caught on a submerged log, so I hopped in to free it. When we finally got the net to the surface, we found it was bursting with giant black piranhas. The waters I'd just been swimming in were truly infested with piranhas!

RAZOR SHARP
I generally try to avoid clichés such as 'razor sharp', but for piranha teeth, it is totally true.

A piranha's teeth are made for quickly cutting into and biting off chunks of flesh from its victim.

A piranha has an excellent sense of smell which helps it find food in the water.

Power

Strength

The grizzly bear is a true powerhouse. Despite weighing half a tonne or more, grizzlies have been clocked moving at 64 kilometres an hour in short bursts, and they can kill with a single swipe of their massive paws. They are only slightly smaller than polar bears, but their habits are very different. Grizzly bears are omnivorous, with most of their diet made up of plant matter. They may gorge on salmon though, when it is plentiful, and they hunt animals as large as moose and bison. Polar bears are rarely seen eating plants and feed entirely on meat. For this reason, polar bears are among the most dangerous animals on Earth to human beings – if we are in their territory, we are potential food.

MANTIS SHRIMP

This colourful crustacean can smash its way through crab shells. Its punch is said to be faster than a .22 bullet and capable of smashing through bulletproof glass!

LEOPARD SEAL

The leopard seal is a massive animal, nearly 4 metres long and weighing up to 450 kilos. It's probably the fiercest of all seals and it hunts other seals as well as penguins.

The carnivorous leopard seal is certainly an animal to be approached with respect. Though leopard seals have usually acted with inquisitive intelligence to human beings, one dragged an Antarctic scientist to her death in 2003.

STRONG STOAT

This slender little creature is powerful for its size. It will kill a rabbit ten times its weight, then drag it away to hide or share with its kits.

Bears are the largest of all land carnivores and armed with huge canine teeth and oversized paws and claws. They have one of the finest senses of smell in the natural world, but comparatively poor eyesight and hearing.

Trapped with grizzlies!

While filming in Alaska we stopped at a waterfall to watch three grizzlies catching salmon. Unfortunately, we were so engrossed that we didn't notice the tide coming in. We were trapped with the bears, and when one tried to get out right over the top of us, we thought we were all goners! Luckily, he decided he was full of salmon and didn't fancy a confrontation, so turned away at the very last minute.

Constrictor Snakes

Reptiles are often described as cold-blooded, although many have blood that is warmer than a human being could endure. The correct term is ectothermic (getting heat from the outside) or poikilothermic (with a dramatically varying body heat). The advantage of getting body warmth from the outside is that the animal doesn't have to be constantly processing food to control its body temperature. Because of this, reptiles can fast for a long time – the record was a Habu pit viper which went for three years without eating, and suffered no obvious ill effects.

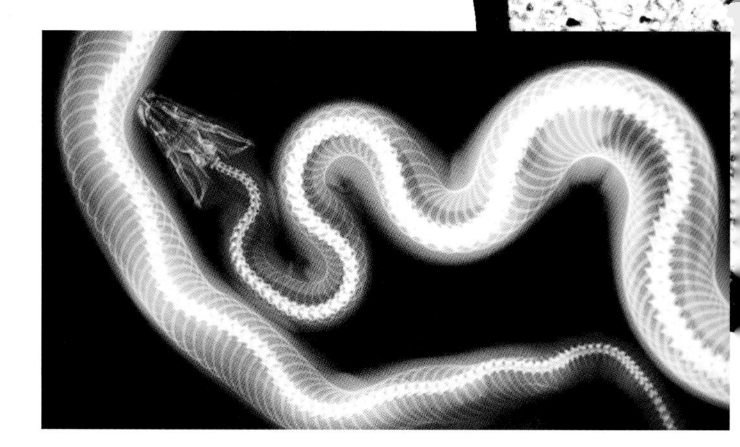

An X-ray of a boa's skeleton shows that its body is mostly made up of vertebrae and ribs. Humans have 12 pairs of ribs. Snakes have 200 or more.

ANACONDA

Really large snakes spend much of their time in water, as it helps to support their weight. The anaconda is easily the heaviest of all snakes and lives in swamps and rivers in South American jungles. Like all constrictors, it kills prey, such as the caiman shown here, by squeezing with its powerful body coils until the prey suffocates. The anaconda can also open its mouth extraordinarily wide to swallow large prey whole.

My boa moment

Don't worry, I put this boa round my neck on purpose to demonstrate its constricting powers! I certainly wouldn't do this with a wild snake any bigger than this one, as it could choke the life out of me. After just a minute or so, even this boa, which was only 2 metres long, had me gasping, purple in the face and genuinely struggling to take a breath.

Powerful body coils for squeezing prey to death.

Boa flickers tongue in and out to sense its surroundings.

Leap

Looking at champion springers, such as the tarsier, the rocket frog and the lynx, it's easy to see what gives them their superlative leap. Their legs are long and comparatively thin, but driven by heavily elastic ligaments and tendons, and powerful muscles built for explosive force. When the animal lands, the tendons absorb the force and can bounce it off again, almost like a rubber ball. However, in most leapers, because the muscles driving the jump are full of fast-twitch muscle fibres, built for sudden quick movements, the animal soon tires and cannot keep leaping for long.

TARSIER
The tarsier hunts in darkness, its huge eyes sucking in every scrap of light and its twitching ears homing in on the tiniest rustle. When it locates a target, it leaps, using its incredibly springy back legs.

A bouncing tarsier

It was late at night. We'd not had much luck with filming tarsiers, and all the crew were desperate to head home to bed, but I was not going to give up. I dragged the grumbling crew around and finally, a couple of hours later, we had the remarkable experience of having a tarsier bounce down to the camera and catch a bush cricket right in front of my face – wonderful!

The rocket frog is a predator and catches insects with its long sticky tongue. Its impressive jump is not part of its hunting strategy but a way of escaping from animals that might fancy making it a meal.

LOOK AT HOW FAR THESE ANIMALS CAN LEAP COMPARED TO THEIR BODY SIZE.

LEAPING LYNX

The lynx is a cat that relies on creeping close to prey and then killing it with one final, decisive pounce. It can leap up into the lower branches of trees, and it's a pretty good climber too.

Climbers

Success in the natural world is all about making the most of opportunities, and predators that can climb have whole new worlds at their disposal. The canopy of a tropical rainforest may contain more species than anywhere else on Earth, and a carnivore that can scamper around in the treetops has an unbelievably rich array of food at its disposal. There are no half measures though. Attempting a kill way above the forest floor is seriously risky – one misjudged leap could send a predator tumbling to its death.

OCELOT

These medium-sized cats of Central and South American forests, are perhaps the New World equivalent of clouded leopards. They are happy hunting birds and monkeys high in the trees.

CLINGING ON

True tree frogs have disk-like pads under each toe, which are kept moist and act like sucker cups. I've seen tree frogs leap out at branches and large leaves, then dangle by just a couple of these sucker cups.

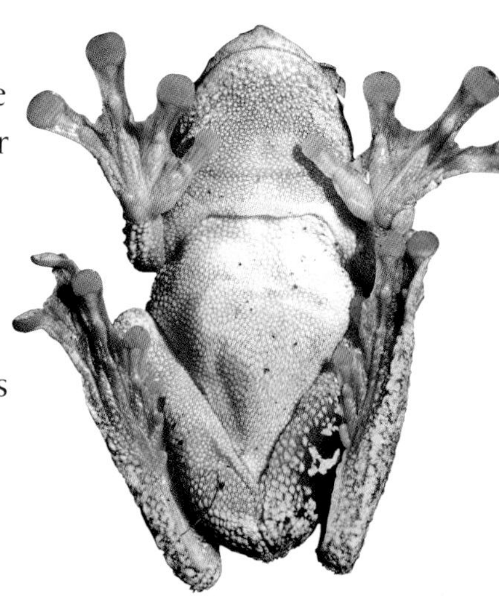

When you look at the underside of a gecko's toes under an electron microscope, you can see successively smaller cusps, branches and tiny hairs. These actually bond to imperfections in surfaces and allow the gecko to grip.

ACROBATIC GECKO
The gecko's ability to scamper over vertical or overhanging surfaces is unmatched among vertebrates.

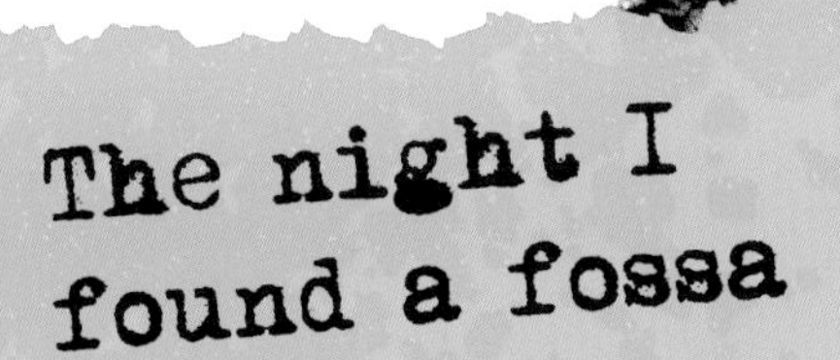

The night I found a fossa

Fossas are incredibly skilful climbers and can run down tree trunks as well as up as they chase prey. I was nervous about our chances of filming a fossa in the wild, as they are rare, nocturnal and extremely secretive. However, two males wandered into our camp the first night, and took to sleeping under our kit hut by day! They seemed to exude a sense of arrogance, of being unafraid of anything and used to being firmly at the top of the Madagascan food chain.

Ferocity

The group of animals perhaps most capable of punching above their weight are the mustelids, such as stoats, weasels, martens, minks and badgers. Stoats leap on the back of larger prey and bite through their spines. Martens and mink are also extremely voracious. However, the most ferocious fiend of all may well be the wolverine, which is not much larger than a European badger but capable of taking down animals as large as moose – truly a predator worthy of respect.

BADGER
The honey badger is known to frighten lions away by its sheer guts and aggression! With its thick fur and a high resistance to snake venoms, it will take on Africa's most venomous snakes.

MARSUPIAL DEVIL
The Tasmanian devil is more or less the marsupial equivalent of a honey badger. It lives on the island of Tasmania and can hunt good-sized prey, but prefers to scavenge for its food.

The devil's big jaws house strong, sharp teeth and it will chomp down more or less anything it can get hold of, including bones.

WOLVERINE
The wolverine has an extraordinarily powerful bite. Its jaws are so strong that they can crush through bone and frozen flesh.

Running with a wolverine

The wolverine I encountered in Alaska had been brought up in captivity, but even so was a terrifying mass of whirling claws and teeth. As we ran across the snows, he snapped at my ankles and occasionally clamped down on my Achilles tendon. This is exactly the same technique as wolverines use to bring down large prey animals such as moose before eating them alive!

Death from the Skies

Birds live life at a faster pace than humans – flying is hard work and makes great demands on the body systems. Fortunately, a bird's circulation and breathing are much more efficient than ours, and the skeleton is extremely light, as any excess weight at all makes flying more difficult. The ability of some birds, such as birds of prey, to travel at great speeds brings with it enormous power. A golden eagle dropping towards its prey at 160 kilometres an hour is demonstrating what is one of the most powerful of all attacks for an animal of its size.

Sharp-tipped beak is used to tear prey apart, not to kill it.

Curved talons are a bird of prey's main weapon.

The day I flew with an eagle

The black eagle is one of the few birds of prey that work together cooperatively to catch their prey. I've seen a male and female flying in tandem to spook a hyrax – also known as a dassie – from its rocky location. I was lucky enough to soar alongside a black eagle in a paraglider, watching as it sought out thermal air currents on which to ride higher. Unfortunately, my landing was not quite as graceful as the eagle's. I crash landed into the only pile of thorn bushes for miles around!

LONGEST WINGSPANS

A barn owl swoops on its prey and kills with needle-sharp talons. The undersides of its toes are ridged for extra grip.

POWERFUL HUNTERS

Harpy eagles are often said to be the most powerful birds of prey – a harpy has been seen flying off with a 7 kilo howler monkey in its talons! When climbing trees in Panama to film harpy eagles, we had to wear police stab vests in case the birds attacked. James, our tree-climbing expert, had a big female smash into him with her talons. The attack was silent and he never heard it coming, but he is certain that without the vest she would have punctured right through his lungs and heart and killed him. An adult female can weigh 9 kilos.

Senses

Super Senses

Echolocation or biosonar is used by whales, dolphins, oilbirds and cave swiftlets, as well as by insect-eating bats. Sounds, mostly too high for the human ear to hear, travel out in waves. They bounce back off the objects and creatures in their path and come back to the sender, allowing it both to navigate and to find potential food.

Recent research suggests that bats actually deconnect the inner ear bones from the eardrum as they make their high-pitched calls – otherwise they would blow out their own hearing!

Extra-large ears
for picking up sounds.

Nose leaf
Some bats have very complex nostrils which they use to transmit sounds. This arrangement is called a nose leaf and acts like a satellite dish to concentrate the sounds.

LISTENING FOR PREY

The bizarre aye-aye uses its stick-thin middle finger to tap on a branch, then listens with its huge bat-like ears for the quality of the echoes. It seems it can sense the presence of large insect larvae beneath the surface of the wood. The aye-aye gnaws at the wood with its huge rodent-like teeth, then hooks its find from its hiding place with its long multi-purpose finger.

ELECTRICAL SENSE

The platypus is the only mammal known to be able to detect electric fields. When diving, its eyes are closed and its ears sealed by a flap of skin, so it is blind and deaf underwater. However, the platypus's soft muzzle contains electrical receptors that can detect the tiny electrical discharges created by the moving muscles of creatures such as crayfish. The platypus can find them from about 10 centimetres away, even if they are concealed in mud.

WATER MONITORS

Water monitors have sharp claws and teeth for attacking prey, but they find their food with the help of their long forked tongue. The lizard flicks its tongue in and out to 'taste' its prey's scent and detect its position.

Up close with a monitor lizard

I've learned to treat monitor lizards with great respect. A colleague, who has caught every notable venomous snake, had his worst injury from being mauled by a medium-sized lace monitor. Their teeth and their attitude combine to make monitors a fearsome prospect, so when a substantial water monitor lizard started nosing around just centimetres from my face, I have to admit to being nervous. He was flickering his long forked tongue into the leaves in search of food, but he also flicked his tongue near my face to taste me too!

Eyesight

From examining the different kinds of eyes found in the animal kingdom, scientists think that eyes have evolved independently 40 or more times. The theory is that the first simple eye may have been little more than an area of dark pigment on a cell. This would have warmed up more quickly than other areas and so allowed an organism to move towards light and warmth. Over millions of years, these cells could well have evolved into retinas.

PERFECT EYES

The cuttlefish eye is probably one of the most perfect of all. It has no 'blind spot' like a human eye does, it can focus forwards and backwards and can see into spectrums of light we cannot perceive. Not surprisingly, the cuttlefish relies on sight for hunting.

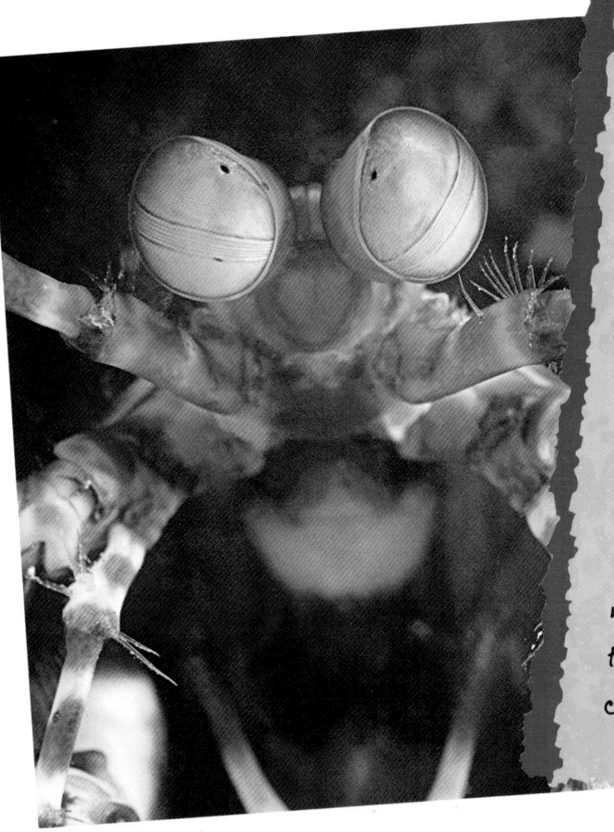

SUPER EYES

Mantis shrimps may have the most complex sight system of any animal. They can perceive well into both UV and infrared spectrums, and while we have three colour channels, they have at least 12.

Eye to eye with mantis shrimps

Filming with mantis shrimps on the sandy bottom of the Bornean sea was one of my true wildlife highlights. I tempted them to lunge out of their burrows with a tiny piece of shrimp, but was unprepared for the speed of their strike, and the sheer size of these prehistoric looking creatures. However, it's their remarkable eyesight that makes them so formidable; they see in colours we simply cannot perceive.

The African fish eagle has superb eyesight. It can even correct for refraction in the water as it swoops down to snatch fish from just below the surface.

An eagle's sight is eight times sharper than ours. It is said to be able to spot a moving hare from 3 kilometres away.

Jumping spiders do not make webs. They creep on their prey, then pounce.

A jumping spider has four pairs of eyes. The two main ones can see in low light and spot movement with great accuracy.

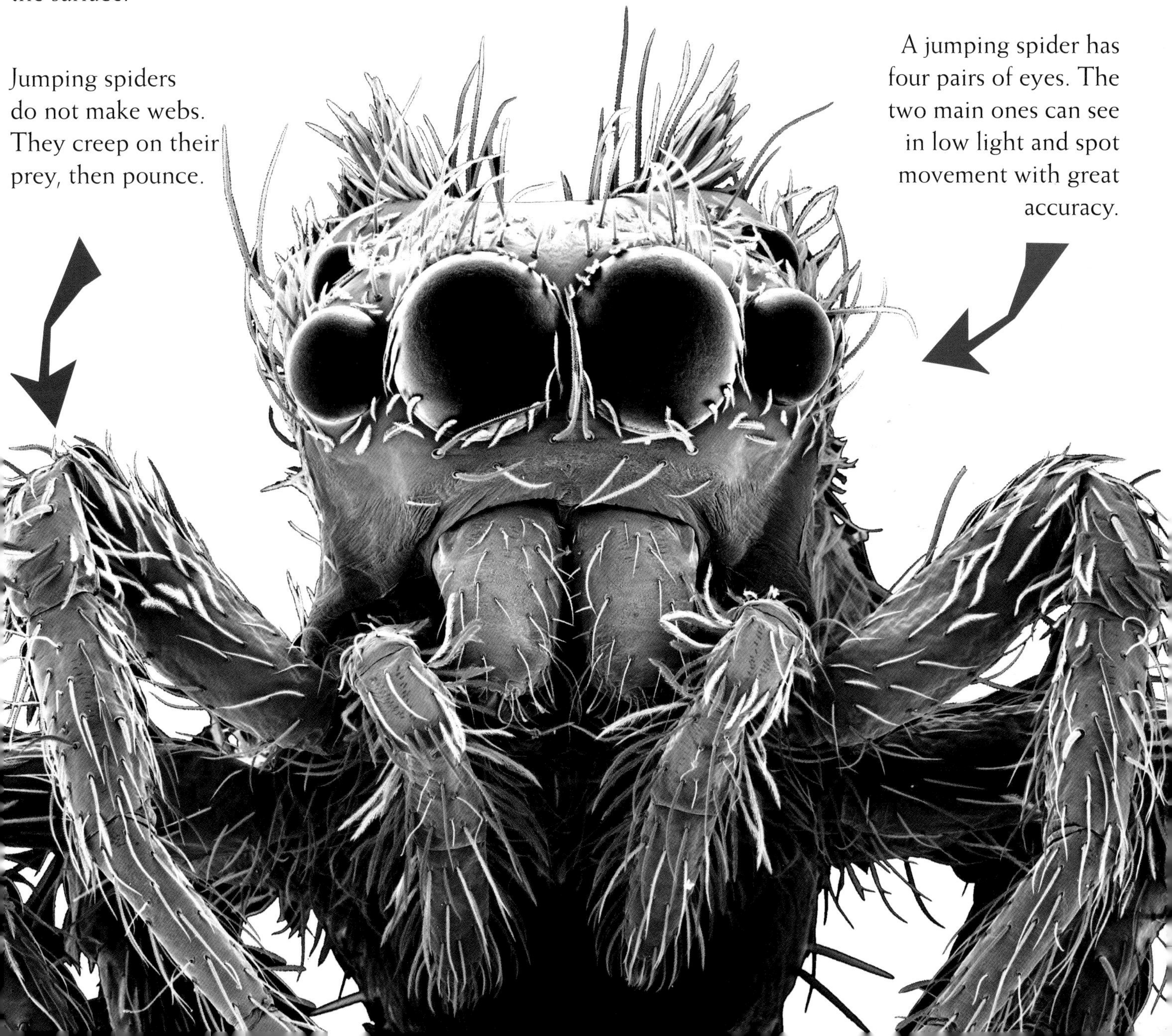

Smell

For many animals, their sense of smell is as important as our sense of sight is to us. Male emperor moths can pick up single molecules of a female's pheromones (smelly sex chemicals) from 10 kilometres away, and some dogs can detect traces of scent that are months old and hopelessly degraded. Many animals that are driven by smell, such as dogs and bears, have long muzzles, with enlarged nasal cavities and a larger surface area of olfactory epithelium – scent receptors.

A dog's long nose allows plenty of room for its smelling equipment – the scent receptors.

The polar bear's nose is one of the few sensitive parts of the body not covered by warm white fur. But it's not true that polar bears cover up their nose when they hunt, as many people believe!

Scented by a bear

Polar bears have been observed in the Arctic walking for 32 kilometres – some sources suggest 64 kilometres!!! – in a dead straight line towards a food source, perhaps a seal colony or a carcass. It's commonly believed that they have actually picked up the scent from all that distance away. While we were in the Arctic trying to film polar bears, we came across a bear who was probably 1.5 kilometres away across the ice. No sooner had we arrived than he lifted his nose to the air, scented us and ambled off in the other direction. This would have been less remarkable if the wind hadn't been blowing away from the bear and towards us!

NATURE'S TOP SMELLERS

BROWN BEAR
The brown bear's brain is a third of the size of ours, but the part of the brain that processes smells is five times bigger. Better than a bloodhound.

POLAR BEAR
This bear can scent a seal carcass from 32 kilometres away.

WOLF
A wolf can detect moose with calves from over 6 kilometres away. Its sense of smell may be more than 1,000 times better than ours.

BLACK-FOOTED ALBATROSS
This bird can smell bacon fat being poured into the sea from 30 kilometres away!

GAMBIAN POUCHED RAT
Their sense of smell is so acute that these rats have been trained to sniff out land mines.

SHARK
Two-thirds of a shark's brain is dedicated to smell; it can sense a few drops of blood in an Olympic swimming pool's worth of water. Lemon sharks can detect a smelly substance that's been diluted ten million times.

This baby lemon shark has been the subject of scientific study and I'm now releasing it back into the wild.

Touch and Hearing

For humans, touch is a sense that's largely focused on our fingertips, but in some animals almost the whole body is freakishly sensitive. In others, the sense of touch is accentuated by having specialised hairs that are connected to nerve endings. These can detect not only direct pressure but also transmitted sound and pressure vibrations. The whiskers on the faces of many mammals such as cats and sea lions have these tactile functions, as do the fine hairs that cover the bodies and legs of many spiders and insects.

Sensitive leg hairs
Touch is a tarantula's most important sense.

HAIRY TARANTULA
The hairiness of a tarantula may be part of the reason many people find them so creepy, but it is actually the source of one of their keenest senses. The hair doesn't insulate, but instead it transmits vibrations in the air and from the ground, giving the spider a picture of movement around it.

Diving with sea lions

Diving with large sea lions can be infinitely more intimidating than diving with sharks, and they don't get any bigger than the Steller sea lion. I dropped in alongside a colony in British Colombia, and the dominant male of the group checked me out as soon as I hit the bottom. These animals weigh a tonne, and have skulls and teeth more impressive than a polar bear. It was truly terrifying stuff. However, when the male lost interest in me, the females all came in to play, biting my fins and even my head! As they came in close, they checked everything out with their whiskers first. Whiskers are clearly the most sensitive part of the sea lion's arsenal.

Soft-edged feathers make less sound in flight

KEEN EARS

The champions of hearing are probably owls. The owl's soft plumage has evolved to allow it to fly silently towards prey. An owl's keen ears lets it hear the movements of prey animals above any wind rush across the wing.

WHISKERS

All seals and sea lions have whiskers. A male Antarctic fur seal was recorded as having whiskers 48 centimetres long. The walrus's whiskers are only 8 centimetres long, the shortest of any pinniped, but it has 300 of them, more than any other species.

Steller sea lions live in the North Pacific.

Intelligence

Intelligence in animals is a really difficult thing to define, particularly because tests of animal brains are entirely on human terms. These tests usually look at animals' ability to learn language and manipulate things with their hands. It's generally believed that the great apes are the brainiest non-human animals. Chimpanzees can learn sign language to communicate with humans, and in the wild they use plants to heal certain ailments. Chimps are also among the very few animals known to use tools to help them find food. They break nuts with rocks, winkle termites from their nests with sticks, and use leaves as sponges. Other creatures, such as whales, also show great brain power.

A near miss!

While filming bubble-netting humpbacks in Alaska, my camerawoman Justine and I were kayaking among the whales as they fed. They seemed to take no notice of us whatsoever. But one morning there was an unearthly silence, the water went totally still, and all the whales disappeared. Then a neat ring of bubbles erupted around us, with our kayak at the very centre. I started backpaddling frantically, while as many as 15 whales surged to the surface and only just missed swallowing us!

WHALE SONAR

Sperm whales have sonar that can not only be used to detect prey, but also in extraordinarily fierce bursts to stun animals. The sonar is focused in a large organ in the head, known as the melon, which is filled with fats.

Teamwork requires communication, organisation and therefore intelligence. A spectacular example of coordinated hunting is the bubble-netting of humpback whales. They swim beneath shoals of herring and exhale a curtain of bubbles from their blowholes, trapping the fish. The whales then rush to the surface, filling their gullets with fish.

CLEVER TRICKS

Here a crow is using a piece of twig to lever grubs out of a fallen tree trunk. Crows will also pick up mussels and winkles and drop them onto rocks so they smash open.

This chimp is using a handful of leaves like a sponge to squeeze water into its mouth.

A stick makes a useful tool for extracting ants from a nest. Chimps fray the ends of the sticks with their teeth so they pick up insects more efficiently.

Te

amwork

Wolves and Hunting Dogs

The canids, or dog family, are some of the most successful of all predators, with representatives on every continent except Antarctica. Until human persecution ravaged their numbers, the grey wolf was the most widespread carnivore on Earth, but now wolves are not common in any part of their former range. Grey wolves are the ancestors of all domestic dogs – in fact, dogs are nowadays generally classified as *Canis lupus familiaris*, a subspecies of the wolf. Unlike most cats, dogs live in social packs, with the most experienced or capable animals taking the lead in hunting and breeding.

DOG PACKS
In the days before stringent hunting, Cape hunting dogs lived in packs of as many as 500 animals. Now any pack over 30 animals is incredibly rare. Grey wolf packs of nearly 40 animals have been reported but are unusual. A pack of 8–12 animals is much more common.

GREY WOLVES
Wolf packs are led by an alpha male and female. Fights and squabbles are common but, though they may look bloodcurdling, attacks within a pack rarely cause damage. After all, an injury to one pack member affects the hunting ability of the group.

Dominant Attack Aggressive

Relaxed Submissive Afraid

The way a wolf holds its tail sends important messages to other wolves.

A wolf's howl can be heard from 10 kilometres away, and is generally used to maintain contact with other members of the pack. It can also help gather support before a hunt, and strengthen social bonds within the group.

The first time I saw a wolf

I will never forget the first time I saw a wolf up close. As someone who is obsessed with reptiles and bugs, it was a real surprise to be so moved by a mammal. There was something entrancing about the wolf, its transfixing eyes, its complex and fascinating social systems, and of course its penetrating, spine-tingling howl. Wolves have been my favourite animals ever since.

Lions

The lion is the world's second largest cat, with a maximum weight in the wild of about 250 kilos. Females, who are smaller and not encumbered by hefty manes, take on most of the hunting duties. The reason why lions have taken to living and hunting in prides is not entirely clear. Scientists have discovered that when working as a pride, lions are, on average, successful in one out of every three hunts. Working alone, they may only succeed in one in five. However, when hunting alone a lion has the proceeds to itself – provided it can defend the food from hyenas or other lions, whereas a pride has to share its kills.

The lion's mane has evolved because of social dynamics. The male's main focus is protecting his pride. The mane makes him look larger and more intimidating – and makes it harder for a challenger to get a hold of his throat. But the mane is a real disadvantage when hunting, and can make the lion overheat if he tries a protracted hunt.

Lions are massive animals and have a heavy skull and bones. Their huge canine teeth enable them to deliver a killer bite to almost any part of their prey, though they often clamp the muzzle, windpipe or spine of their victims. Cubs learn to kill by watching their mothers and practising their skills on small animals.

Too close!

The closest I've ever been to a wild lion was when tracking a group in order to dart one and put on a radio collar. We followed the lionesses at night in an open-backed truck, and all of a sudden realised one of them was stalking us and was just metres behind the truck. It was incredibly frightening. She could have been in the truck in a second.

Prides of lions are usually dominated by one or two adult males, and there may be several lower-ranking males, as well as related adult females and their offspring. An average pride contains 2 males, 9 females and 15 young.

The pride work together to hunt and raise their cubs. The size of the pride usually depends on the abundance of prey; in some areas where food is plentiful, a pride might number 40 animals. Territories can cover over 260 square kilometres, and may contain many prides, as well as pairs of nomadic lions, and individuals.

Orca

Though I have always been loath to name any one animal as the world's most perfect predator, my heart says the orca, or killer whale, is the champion. Orcas are known to have killed great white sharks and some of the largest whales, and they are certainly one of the most intelligent of all animals. They have complex social structures and ways of communicating that we are not even close to understanding.

Watching orcas

Orcas are the largest species of dolphin. I've caught fleeting glimpses of them over the years, but they've always been travelling and at their cruising speed of 35 kmh they disappear remarkably quickly. They are one of the most impressive creatures on our planet, so I never gave up hope.

We were in Canada, trying to film giant Pacific octopus underwater and failing dismally. I was underwater, shivering away and very frustrated, when the guys up on the boat called down to me over the underwater comms, saying: 'Steve, there's a whole orca pod swimming right past. Get up here as fast as you can!' I was back up on deck and out in my kayak in a matter of minutes! Orcas, like all dolphins, divide their time between feeding, travelling and socialising.

Luckily for me, this pod was in play mode! They were breaching out of the water around me and one even swam the length of my kayak. She was so close I could have run my hand down her flank. As she passed, she rolled onto her side to look me in the eye, obviously just wanting to check me out.

I thought this had been one of the most exciting experiences of my life, but the next day things got even better. This time I was in the kayak on my own while the crew were having lunch, when suddenly a vast male orca scythed up alongside me. His dorsal fin was as tall as I am. The pod were in a totally different mood, and I soon saw why. They were chasing a fully grown Steller sea lion. They drove him underwater by breaching on top of him, beating him into submission, before tossing him right out of the water with their tails. Male Steller sea lions can weigh a tonne or more!

Finally, when the sea lion was so weak it seemed that the orcas must surely finish him off, they allowed their calves to practise their hunting skills on him, just like a cheetah bringing back a baby gazelle alive so her cubs can learn to kill. It was an incredible sight.

The biggest male orca ever recorded was 9.8 metres long. Orcas live and hunt in pods – family groups of up to 40 or so animals. The pod members keep in constant touch, with lots of different sounds.

Extrem

ophiles

Cold

Extremophiles are animals that love living in extreme environments, and the Arctic is one of the most challenging for any animal to live in. Temperatures rarely rise above freezing, even in summer, and in winter there are months of constant darkness. For warm-blooded birds and mammals, the most important adaptation is obviously insulation, achieved in some animals by thick fat called blubber under the skin, and in the birds by layers of densely packed feathers.

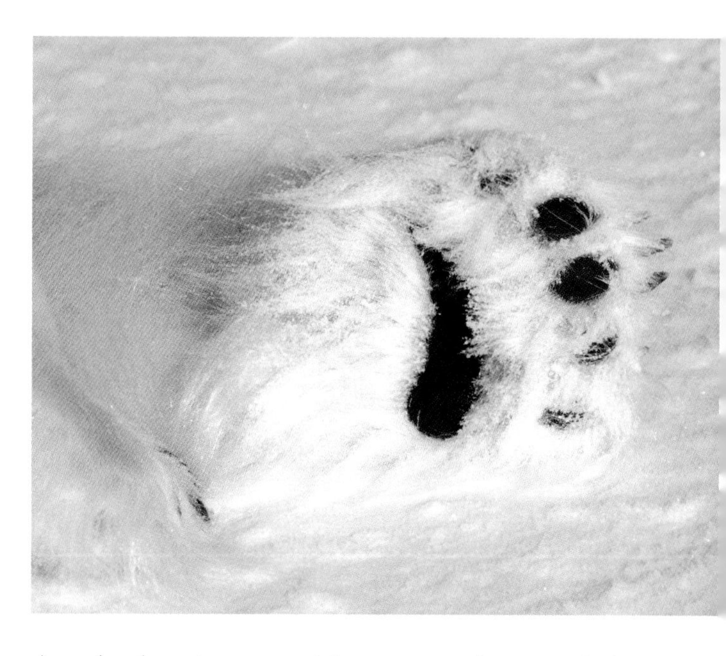

A polar bear's paw is like a snowshoe, and the pads are well protected with plenty of hair.

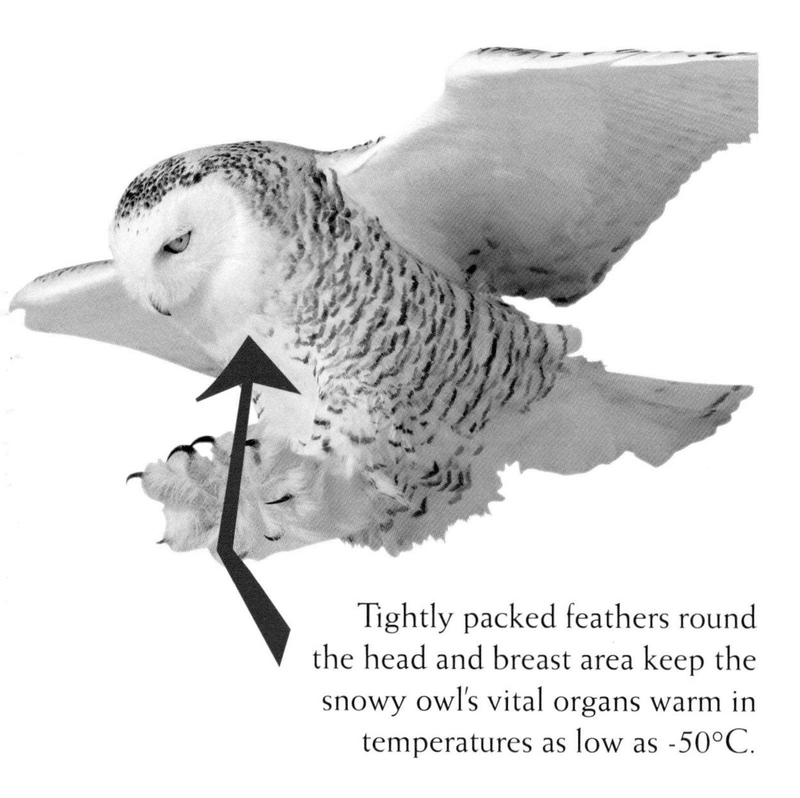

Tightly packed feathers round the head and breast area keep the snowy owl's vital organs warm in temperatures as low as -50°C.

ICE OWL

A snowy owl's dense feathers give it extra weight, which allows it to punch right down through the snow crust to catch lemmings scuttling around beneath. It locates these with its incredible hearing.

POLAR BEAR

A polar bear's fur does such a good job of keeping it warm that very little of its body heat escapes. If you look at a bear through a thermal imaging camera which 'sees' the heat given off (below). Only the uncovered areas around the eyes and nose show up; the rest of the body is efficiently insulated by the fur.

An adult male polar bear has a stomach capacity of 68 kilos. Polar bears are known to kill walruses and beluga whales, both of which weigh a tonne or more.

Tracks in the snow

While tracking polar bears on Alaska's north shore, I found some really recent tracks. Running alongside them were a perfect set of much smaller dog-like prints, from an Arctic fox. Sometimes these sneaky foxes will track a polar bear for weeks on end, hoping to pick up the scraps when the more powerful animal makes a kill.

Heat

Heat is a serious and often lethal challenge to life. Mammals need to ensure their core body temperature stays very close to 37 degrees, as an increase to this can lead to heat stress and heat stroke. Animals that live in extremely hot conditions need to have both physical and behavioural adaptations if they are to survive. Strategies include hiding away in burrows during the day and emerging at night, or shuttling in and out of the shade.

My run in the desert

I've done many expeditions in the desert. At night it's one of the most exciting environments on Earth, but in the daytime it can be ferocious. When running the Marathon Des Sables, 243 kilometres across the Sahara desert, I got heat stroke – hyperthermia. This is the opposite of hypothermia, meaning my core body temperature went up several degrees. I started hallucinating and felt as though I had a furnace inside me. If it hadn't been at the end of the day, with the nighttime cold approaching, I would probably have died!

DESERT FOX
The smallest of the foxes, the fennec fox has the biggest ears. These help it lose heat from its body and stay cool in its desert home.

Like many other reptiles, the sidewinder takes advantage of the fact that just beneath the surface of the sand, the temperature may be 15 or 20 degrees cooler. While most reptiles simply lie low for the day, and emerge at night to feed, the sidewinder uses the sand as cover to make good its attack, keeping just its eyes and nostrils above the surface.

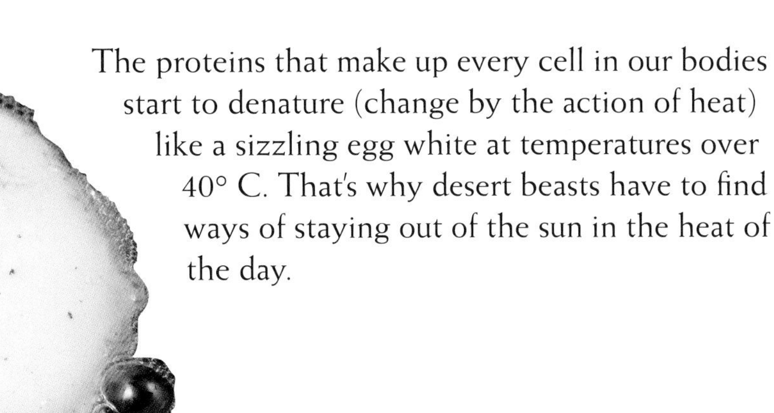

The proteins that make up every cell in our bodies start to denature (change by the action of heat) like a sizzling egg white at temperatures over 40° C. That's why desert beasts have to find ways of staying out of the sun in the heat of the day.

These little geckos live in the Namib desert. They use the webbing between their toes to prevent them sinking into the sand as they walk.

Web-footed gecko
Like most geckos, this lizard has no eyelids, so has to lick its eyes to keep them clean and moist.

Altitude

High altitude poses three main challenges for animals: temperature, climate and oxygen deficiency. For every 200 metres of altitude, the temperature drops by one degree. In addition, mountains attract rain, snow and high winds, all of which can make life thoroughly uncomfortable. Animal life is rare above a certain height, because the air is so thin. At the summit of Everest, there is only a third of the usual amount of oxygen available with every breath. Mountaineers reach the top after acclimatising, but if you were transported there instantly from sea level, you would live for no more than a couple of minutes before dying from lack of oxygen.

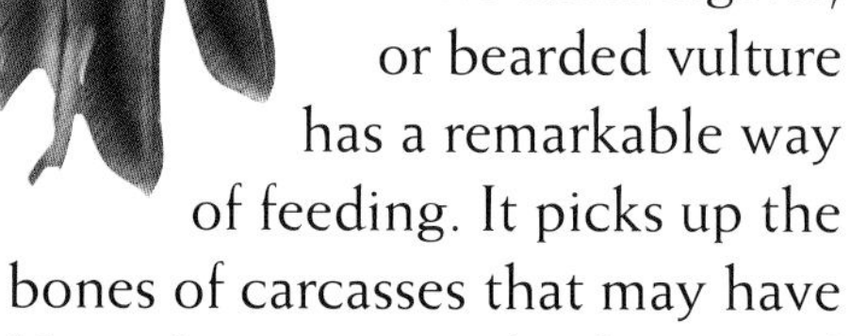

HIGH FLIER

The lammergeier, or bearded vulture has a remarkable way of feeding. It picks up the bones of carcasses that may have been stripped by other species of vultures and carries them high above rocks. The lammergeier then drops them and swoops down to eat the marrow from the splintered remains.

WHERE EAGLES SOAR

The golden eagle is one of the largest and most powerful of all birds and generally lives in wild, mountainous landscapes. Golden eagles have been seen flying straight at sheep, mountain goats and red deer on steep cliffsides, driving them to panic and fall.

Eagles kill smaller prey outright in a stoop that may be as fast as 160 km/h, and ends with the bird driving its long talons into the prey's chest and vital organs.

Perhaps no animal represents the high Himalayas more than the snow leopard, one of the most beautiful and elusive creatures on Earth. Its dense fur coat allows it to live well above the snow line – in fact, snow leopards rarely come down to below about 4,000 metres above sea level.

My night with a snow leopard

When trying to film snow leopards in the Himalayan Kingdom of Bhutan, I sat alone at night next to a carcass that had been killed the night before by a large male snow leopard. After a few hours of sitting in total silence, I heard the leopard just metres away. He could smell me, and gave a vicious snarl chilling me to the bone, then sprinted off knocking stones down alongside me. I have to admit I was absolutely terrified, which isn't surprising really; the sheep carcass I was sitting beside looked as if it had been sawn in half with a chainsaw.

Deep Sea

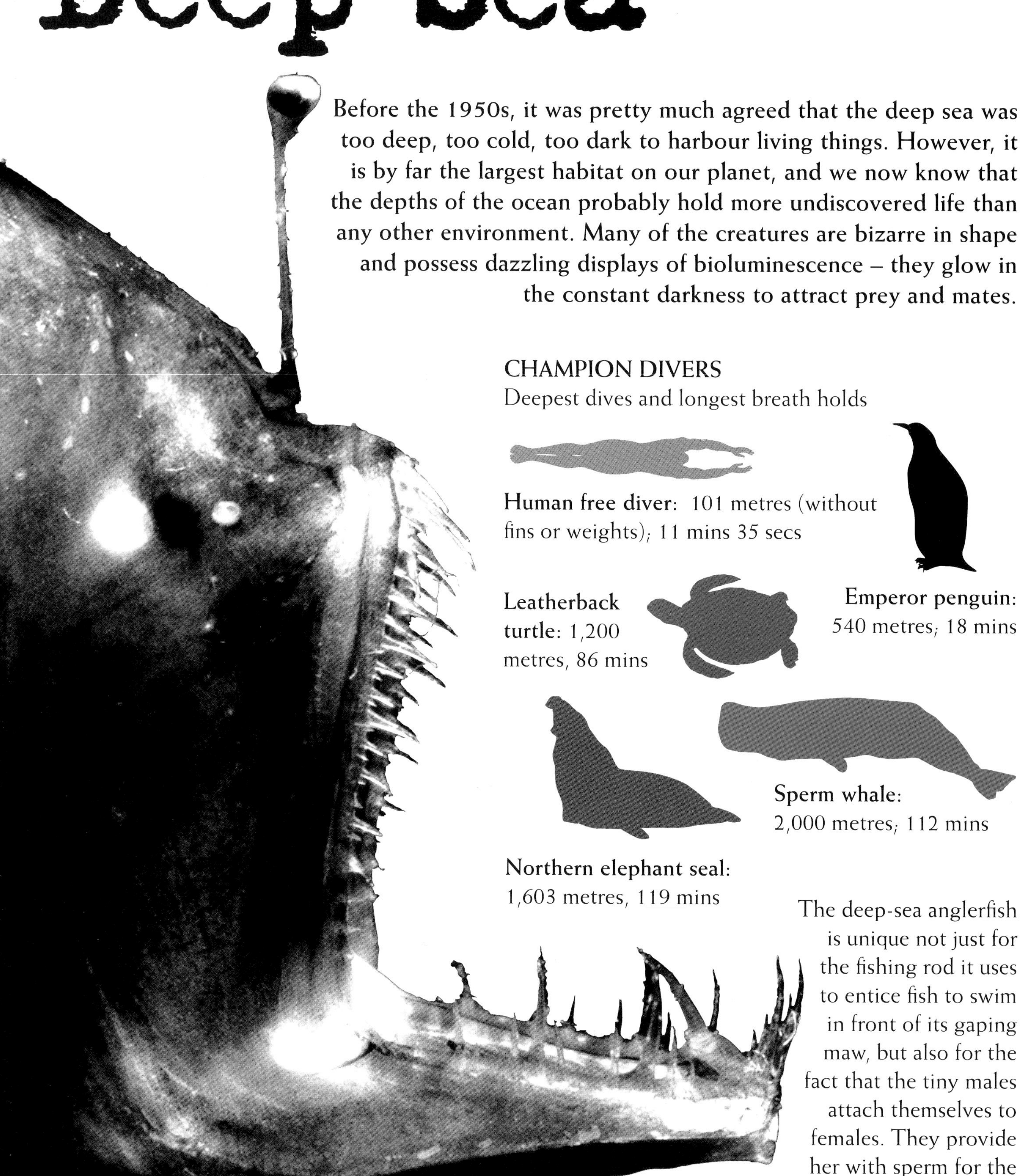

Before the 1950s, it was pretty much agreed that the deep sea was too deep, too cold, too dark to harbour living things. However, it is by far the largest habitat on our planet, and we now know that the depths of the ocean probably hold more undiscovered life than any other environment. Many of the creatures are bizarre in shape and possess dazzling displays of bioluminescence – they glow in the constant darkness to attract prey and mates.

CHAMPION DIVERS

Deepest dives and longest breath holds

Human free diver: 101 metres (without fins or weights); 11 mins 35 secs

Leatherback turtle: 1,200 metres, 86 mins

Emperor penguin: 540 metres; 18 mins

Sperm whale: 2,000 metres; 112 mins

Northern elephant seal: 1,603 metres, 119 mins

The deep-sea anglerfish is unique not just for the fishing rod it uses to entice fish to swim in front of its gaping maw, but also for the fact that the tiny males attach themselves to females. They provide her with sperm for the rest of her life, while they get a free ride!

TENTACLE TEETH

The sucker cups of the Humboldt squid are lined with hundreds of teeth which look just like those you'd expect to find in a piranha's mouth!

MONSTER OF THE DEEP

A ferocious, fast-moving hunter, the Humboldt squid is 2 metres long and preys on fish, shellfish and other squid.

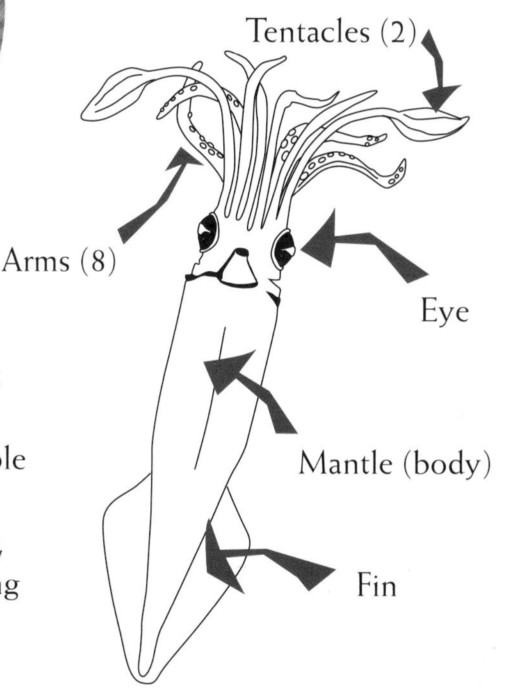

The beak of a Humboldt squid is rock-hard at the biting tip and more supple towards the outsides, providing great flexibility, as well as frightening biting power.

Down deep with a Humboldt squid

This was possibly the most frightening underwater encounter I've had. Shortly after nightfall, these deep sea monsters come into shallower waters to feed, and can be enticed into the human zone. We not only had to wear chainmail suits to avoid being bitten and torn apart by their sucker teeth, but we also had to be on steel safety cables to stop the squid dragging us down into the depths! Having these formidable beasts approach from out of the gloom was truly terrifying.

Endangered

It's a tragedy that we need to talk about endangered animals in this book. Unfortunately, there is no doubt that we are the most deadly, and by far the scariest, animals on Earth. No other animal is so effective at both destroying the planet we rely on for our existence and wiping out the creatures we share it with. Because of us, many of the most spectacular animals in the world are on the brink of extinction. For example, an estimated 100 million sharks are taken from the world's seas every year – most of them killed for shark's fin soup. After their fins have been removed, the sharks are thrown back into the sea to die a slow death. Here are some of the reasons why the world's animals are in trouble.

Habitat loss: The only way to preserve species is to maintain the areas they live in. When an acre of rainforest is cut down, millions of animals are left without anywhere to live. Monoculture (which is the growing of one kind of crop on land that would once have harboured thousands of species of plants) results in plummeting biodiversity. Insect numbers dwindle, the animals that feed on them suffer too, and the whole food chain falls apart.

Global fishing crisis: Some kinds of fishing are so destructive that our seas will never recover from them and mean that many species will be extinct before we even know of their existence. It is encouraging, though, that when 'no-take zones' are set up, localised fish stocks recover spectacularly.

Climate change: The release of carbon gases into the atmosphere is causing an overall warming of our climate. The effects on wildlife are too numerous for us to even understand, but could result in the death of the world's coral reefs, and huge areas turning to desert.

Poaching: Many animals have body parts that are valued by certain cultures. Ivory is used for carvings, rhino horn for dagger handles, tiger body parts for traditional medicines. Animals are targeted by gangs who can earn a year's wages for a single kill.

Bushmeat: Overpopulation is the greatest problem facing our planet. As the population spirals out of control, people move into the few sanctuaries that still remain for wildlife. Animals become a source of food, and are hunted unsustainably.

Illegal pet trade: More and more people want to own certain kinds of wild animals, and rare animals command high prices. What people don't see is how many animals have to die for one creature to make it to market in another country.

The fate of our world's wildlife is in your hands. Join conservation organisations. Learn as much about the issues facing our wild animals and environments as you can. The only limit to what you can do to preserve wildlife is your own ambition and creativity. **I'M COUNTING ON YOU!**

Steve

Index

albatross 63, 71
anaconda 54
anglerfish 92
ants 27, 33, 40
arachnids 19
aye-aye 67

barracuda 14, 15
bats 44, 45, 66
beaks 20, 62, 93
bears 44, 52–3, 70–1, 86–7
birds of prey 20-21, 62, 63, 69, 90
boa constrictor 54, 55

cats 44
centipede 41
chameleon 28–9
cheetah 10–11, 83
chimpanzee 74, 75
claws 21
cobra 38–9
coral reefs 30
crocodile 42, 43, 44
crow 75
cuttlefish 28, 68

death adder 12, 13
deep sea monsters 92–3
deer bot fly 18
dogs 44, 70, 78–79
dolphins 16, 17, 66, 82–83
dragonfly 18, 19
ducks 20

eagles 62, 63, 69, 90
ears 56, 66, 67, 73, 88
eggs 27
electrical fields 67
elephant seal 92
eyes 20, 26, 27, 29, 68–9, 89

fangs 33, 34, 38, 39, 40
feathers 73, 86
feet 29, 86, 89
fish 14–15, 30–1, 40–1, 48–9
fleas 57
flies 18
fossa 59
foxes 87, 88
frogfish 31
frogs 24, 41, 56, 57, 58
fur 16, 24, 60, 70, 86, 91

gecko 28–9, 89
geese 21
gentoo penguin 17
gila monster 41
glossy racer 12, 13

hearing 66–7, 72–3
honey badger 60
horseflies 18
hunting dogs 78
hyena 42

insects 18, 19, 24
see also named types

jaguar 11, 24
jaws 18, 34, 43, 46, 60, 61

kangaroos 57

lammergeier 90
leech 44
leopards 11, 24, 25, 57, 91
lionfish 41
lion 24, 42, 80–1
lizards 24, 41, 44, 67
lynx 56, 57

mamba 12, 38
mantids 26–7
mantis shrimp 52, 68
margay 57
marlin 14, 15
monitor lizard 44, 67
moths 18, 70
mouths 15, 26, 31, 33, 54

noses 66, 70

ocelot 58
octopus 28, 33, 40
oilbird 66
orca 82–3
owls 63, 73, 86

penguins 16, 17, 92
peregrine falcon 20–1
pigeons 20
piranha 48–9
platypus 67
poison 40, 41
polar bear 44, 52, 70–1, 86–7
pufferfish 40
puma 57

rats 71
rays 46

sabre-toothed kerasin 44
sailfish 14, 15
scorpions 40, 41
sea dragons 30
sea lions 16, 72, 73, 83
seals 16, 17, 52, 73
sharks 14, 15, 28, 31, 46–7, 71
skulls 11, 38
smell 49, 70–1

snakes 12–13, 32–3, 38–9, 40–1, 44, 54–5, 88
solifugid 19
spiders 33, 34–5, 41, 69, 72
springtail 57
squid 28, 93
stoat 52
stork 63
swiftlet 66

tails 15, 17, 29
taipan 38
talons 21, 62, 63, 90
tarantula 72
tarsier 56
Tasmanian devil 60
teeth 44, 45, 46, 47, 49, 93
see also fangs
tiger 24, 25, 42
tongues 28, 55, 56, 67
tuna 15
turtles 43, 92

vampire bat 44, 45
venom 38, 40, 41
viper 39, 40, 41, 54

walrus 73
wasps 27, 33
whales 16, 66, 74, 82, 83, 92
whiskers 16, 72, 73
wings 18, 19, 20, 21, 63
wobbegong 31
wolverine 60, 61
wolves 71, 78–9

Picture Credits

ARDEA
10 John Daniels; 11 M. Watson; 15 Mark Spencer; 16 Kurt Amsler; 17 M Watson; 20 Jim Zipp; 21 Jim Zipp; 21 Bill Coster; 24 Francois Gohier; 25 Duncan Usher; 27 Andy Teare; 29 Thomas Marent; 36 Kenneth W. Fink; 40 Thomas Marent; 40 Mary Clay; 42 Yann Arthus-Bertrand; 44 Adrian Warren; 45 Nick Gordon; 48 M. Watson; 48 M. Watson; 52 Don Hadden; 53 Auscape; 56 Auscape; 56 Andrey Zvoznnikov; 57 Duncan Usher; 58 Adrian Warren; 58 Auscape; 62 John Mason; 63 Kenneth W. Fink; 66 John Mason; 67 Nick Gordon; 67 Auscape; 69 David Spears; 70 James Marchington; 72 Andy Teare; 73 Rolf Kopfle; 73 M. Watson; 74 Francois Gohier; 75 Adrian Warren; 75 Jean Paul Ferero; 78 Duncan Usher; 78 Chris Harvey; 79 M. Watson; 80 Ferrero Labat: 80 Yann Arthus-Bertrand; 83 Tom and Pat Leeson; 86 Jim Zipp; 87 Tom and Pat Leeson; 89 Karl Terblanche; 90 Jon Cancalosi; 91 John Daniels

NHPA
7 Mark Bowler; 12 Bill Love; 18 Stephen Dalton; 18 James Carmichael Jr; 24 Nick Garbutt; 26 Stephen Dalton; 27 James Carmichael Jr; 32 Mark O'Shea; 33 James Carmichael Jr 35 Nick Garbutt; 46 Ocean Images; 47 Franco Banfi; 52 Stephen Dalton; 53 Rich Kirchner/Photoshot; 54 Bill Love; 54 Martin Wendler; 55 Jany Sauvanet; 56 Biosphoto/Alain Compost; 59 Martin Harvey; 60 Gerry Pearce; 60 Steve and Ann Toon; 61 Mike Hill; 63 Manfred Danegger; 66 A.N.T. Photolibrary; 68 Burt Jones & Maurine Shimlock; 70 Photoshot; 82 Oceans Images/Photoshot; 90 Jordi Bas Casas;

NPL
33 Constantinos Petrinos

OXFORD SCIENTIFIC
17 Paulo de Oliveira/Photolibrary; 68 Obbet Tobias Bernard/photolibrary

SCIENCE PHOTO LIBRARY
12 Karl H Switak; 14 Eye of Science; 15 Richard Herrmann/Visuals Unlimited; 17 David Fleetham/Visuals Unlimited; 18 ; 26 Ray Coleman; 26 Paul Harcourt; 28 Patrick Landmann; 29 Georgette Douwma; 30 Matthew Oldfield; 30 Paul Zahl; 32 Gerald and Cynthia Merker; 32 Suzanne L. Collins; 33 Michael McCoy; 33 Karl H. Switak; 33 Matthew Oldfield; 34 Dr John Brackenbury; 35 Anthony Mercieca; 36 Peter B. Kaplan; 42 Sinclair Stamers; 43 Tom McHugh; 46 Georgette Douwma; 49 Dante Fenolio; 62 Dr. Marazzi; 75 Tom McHugh; 81 Art Wolfe; 86 Dan Guravich; 86 T-Service; 88 Mark Phillips; 89 Solvin Zanki/Visuals Unlimited; 92 Dante Fenolio; 93 Louise Murray

SCUBAZOO
31 Jason Isley/Scubazoo; 41 Adam Broadbent/Scubazoo; ; 82 Simon Enderby/Scubazoo

Additional photography by:
Nick Allison; Steve Backshall; Giles Badger; Charlie Bingham; James Brickell; Kirstine Davison; Rosie Gloyns; Emma Jones; Tim Martin; Rowan Musgrave; Johnny Rogers; Nikki Waldron; Adam White